Understanding the Brain

TOWARDS A NEW LEARNING SCIENCE

ORGANISATION FOR ECONOMIC CO-OPERATION AND DEVELOPMENT

ORGANISATION FOR ECONOMIC CO-OPERATION AND DEVELOPMENT

Pursuant to Article 1 of the Convention signed in Paris on 14th December 1960, and which came into force on 30th September 1961, the Organisation for Economic Co-operation and Development (OECD) shall promote policies designed:

– to achieve the highest sustainable economic growth and employment and a rising standard of living in Member countries, while maintaining financial stability, and thus to contribute to the development of the world economy;

– to contribute to sound economic expansion in Member as well as non-member countries in the process of economic development; and

– to contribute to the expansion of world trade on a multilateral, non-discriminatory basis in accordance with international obligations.

The original Member countries of the OECD are Austria, Belgium, Canada, Denmark, France, Germany, Greece, Iceland, Ireland, Italy, Luxembourg, the Netherlands, Norway, Portugal, Spain, Sweden, Switzerland, Turkey, the United Kingdom and the United States. The following countries became Members subsequently through accession at the dates indicated hereafter: Japan (28th April 1964), Finland (28th January 1969), Australia (7th June 1971), New Zealand (29th May 1973), Mexico (18th May 1994), the Czech Republic (21st December 1995), Hungary (7th May 1996), Poland (22nd November 1996), Korea (12th December 1996) and the Slovak Republic (14th December 2000). The Commission of the European Communities takes part in the work of the OECD (Article 13 of the OECD Convention).

Publié en français sous le titre :
Comprendre le cerveau
VERS UNE NOUVELLE SCIENCE DE L'APPRENTISSAGE

Foreword

The project on "Learning Sciences and Brain Research" was launched by the OECD's CERI in 1999. The purpose of this novel project was to encourage collaboration between learning sciences and brain research on the one hand, and researchers and policy-makers on the other hand. The CERI Governing Board recognised this as a difficult and challenging task, but with a high potential pay-off. It was particularly agreed that the project had excellent potential for better understanding learning processes over the lifecycle and that a number of important ethical issues had to be addressed in this framework. Together, these potentials and concerns highlighted the need for dialogue between the different stakeholders.

Brain research only starts to find application in the learning field, despite the remarkable progress in fundamental research in the last decade. The number of discoveries from brain research that have been exploited by the learning sciences is still slim, perhaps due to the fact that there have historically been few direct contacts between brain and learning scientists, and little consensus on the potential applications of brain research to learning science. But there are various reasons for creating more bridges between the two research communities. For instance, new findings about the brain's plasticity to learn anew over the individual's lifecycle have been made, and new technologies of non-invasive brain scanning and imaging are opening up totally new methods of work for research. By bringing the two research communities closer as part of their work, a likelihood of making more value-added discoveries has been pointed out.

This publication aims to provide a summary introduction to what is now known, what is likely to be revealed shortly and what may ultimately be knowable. It is intended to be accessible to non-specialists and it therefore seeks to avoid exclusive language. Its content derives from the three OECD fora organised respectively in New York City in June 2000 ("Brain Mechanisms and Early Learning"), in Granada in February 2001 ("Brain Mechanisms and Youth Learning"), in Tokyo in April 2001 ("Brain Mechanisms and Learning in Ageing"); it also presents the next steps of the project.

Essential financial and substantive support was provided from the beginning by the National Science Foundation (Directorate of Research, Evaluation and

Communication/Education Division), United States; the Lifelong Learning Foundation, United Kingdom; the City of Granada, Spain; the Japanese Ministry of Education, Culture, Sports, Science, and Technology; essential scientific, financial and organisational support was provided by the Sackler Institute, United States; the University of Granada, Spain; and the RIKEN Brain Science Institute, Japan.

Parts I and III of this book were drafted by Sir Christopher Ball and Part II was drafted by Anthony E. Kelly with the assistance of the OECD Secretariat; the book was partially or completely reviewed by Christopher Brookes, Stanislas Dehaene, Hideaki Koizumi, Stephen Kosslyn, Bruce McCandliss, Michael Posner and Emile Servan-Schreiber. Within the Secretariat, Jarl Bengtsson took the initiative of launching this project and provided strategic and critical support throughout; Vanessa Christoph provided logistical support; the project was co-ordinated by Bruno della Chiesa. This book is published on the responsibility of the Secretary-General of the OECD.

Table of Contents

Acknowledgements

The Secretariat is grateful to the following, who took part in some or all of the fora, and who contributed either in making these fora possible and successful, or in helping the OECD Secretariat to produce this publication, or both:

Sir Christopher Ball, Chancellor, University of Derby (UK); **Richard Bartholomew,** Department for Education and Skills (England); **Jean-Daniel Brèque,** Translator (France); **Christopher Brookes,** Director, The Lifelong Learning Foundation (UK); **John Bruer,** President, James S. McDonnell Foundation (USA); **Stanislas Dehaene,** Research Director, INSERM (France); **Juan Gallo,** Director General de Gabinete, Junta de Andalucía (Spain), and his team; **Eric Hamilton,** Director, Research, Evaluation and Communication/Education Division, National Science Foundation (USA), and his team; **Masao Ito,** Director, RIKEN Brain Science Institute (Japan), and his team; **Anthony E. Kelly,** Professor, Graduate School of Education, George Mason University (USA); **Hideaki Koizumi,** Senior Chief Scientist, Advanced Research Laboratory, Hitachi Ltd. (Japan); **Stephen Kosslyn,** Professor, Department of Psychology, Harvard University (USA); **Bruce McCandliss,** Assistant Professor of Psychology, Sackler Institute, Weill Medical College of Cornell University (USA); **José Moratalla,** Mayor of Granada (Spain), and his team; **Michael Posner,** Professor of Psychology in Psychiatry, Director, Sackler Institute, Weill Medical College of Cornell University (USA); **Teiichi Sato,** Director General, Japan Society for the Promotion of Science and Advisor to the Minister of Education, Culture, Sports, Science, and Technology (Japan); **Emile Servan-Schreiber,** International Consultant (France); **Pio Tudela,** Professor of Psychology, University of Granada (Spain); **Kenneth Whang,** Program Official, National Science Foundation (USA).

Furthermore, the Secretariat would like to express its posthumous gratitude to Rodney Cocking, Program Director, Developmental and Learning Sciences, National Science Foundation (USA). We shall miss him.

Introduction

This book would not have been worth writing a generation ago – and will not be worth recalling a generation from now. But it is timely and relevant today. People living today are fortunate enough to be witness to the accelerating rate of development in the science of the brain and the understanding of human learning. This is a kind of "progress report" on a fast-moving subject, or rather several subjects. In presenting a collaborative and trans-disciplinary account of "learning and the brain" the OECD-CERI initiative attempts to bring together several disciplines to see what they can give to, and gain from, one another.

The aims of this publication are threefold:

- to report on and further develop a creative dialogue between several disciplines and interests (cognitive neuroscience, psychology, education, health, and policy);
- to discover what insights cognitive neuroscience might offer to education and educational policy and vice-versa; and
- to identify questions and issues in the understanding of human learning where education needs help from other disciplines.

Education is not an autonomous discipline. Like medicine or architecture it relies on other disciplines for its theoretical foundation. But, unlike architecture or medicine, education is still in a primitive stage of development. It is an art, not a science.

Consider the following account:

"'The distinctive feature of medical education today is the thoroughness with which theoretical and scientific knowledge are fused with what experience teaches in the practical responsibility of taking care of human beings…' Can the same yet be claimed for teacher-training? Experience of the practical responsibility of teaching young people or adults reveals the over-riding importance of motivation, confidence and a good example of success. With these, learning rarely fails; without them, it rarely succeeds. These, and similar, simple observations drawn from the practical experience of teaching are not as yet underpinned by a secure basis of scientific and theoretical knowledge. The science of learning, a branch of human psychology, is still in its infancy. The theory of learning is pre-scientific – in the sense that it lacks as yet either predictive or explanatory power. We do not understand sufficiently well how children and adults learn to dare to offer an educational or training guarantee. The science of education is in its Linnaean phase, drawing up lists of examples of successful learning, clarifying and sorting effective teaching practices; but it still awaits its Darwin with a powerful explanatory theory of learning."[1]

Education today is a pre-scientific discipline, reliant upon psychology (philosophy, sociology etc.) for its theoretical foundation. This book explores the possibility that cognitive neuroscience might in due course offer a sounder basis for the understanding of learning and the practice of teaching. Some think that this may be a bridge too far at present.[2] It certainly has been in the past, but will it be in the future? We shall see. In any event, it is probably better to be faulted for anticipating the gun than missing it altogether.

It is commonplace to claim that the understanding of the human brain is the last frontier for science. No doubt there will always be a fresh horizon for science to explore as we make intellectual progress. Nonetheless the unravelling of the brain's complexity will be a major step on that journey. It appears that science is on the threshold of substantial advances in the understanding of the brain. This book aims to provide a summary introduction to what is now known, what is likely to be revealed shortly and what may ultimately be knowable. But the best that can be done is to provide no more than a still picture from a fast-moving film.

The science of teaching and learning may indeed be in its infancy, but it too is developing rapidly. A number of factors suggest that the *status quo* may be unsustainable: these include the relative failure of the great educational project of the late nineteenth and twentieth centuries, the impending impact of the new learning technologies, and (of course) the advance of cognitive neuroscience. For more than a century, one in six of young people[3] (and adults reflecting on their childhood) have reported that they "hated school"; a similar proportion have failed to master the elements of literacy and numeracy successfully enough to be securely employable; a similar proportion have played truant from school, disrupted classes or quietly withdrawn their attention from lessons. Successive governments in many nations have made various attempts to improve the situation. But perhaps this is a problem that can't be fixed? Maybe traditional education as we know it inevitably offends one in six pupils? Possibly the classroom model of learning is "brain-unfriendly"?

1. Ball, C. (1991), *Learning Pays*, RSA, London.
2. As Dr. John Bruer has cogently argued. See Chapter 4.
3. This figure comes from the UK; however, according to the first results of the recent PISA study conducted by the OECD, the situation could be even worse throughout the developed countries [see: *www.pisa.oecd.org* and OECD (2001), *Knowledge and Skills for Life – First results from* PISA 2000, Tables 4.1 and 4.2, pp. 265-266].

Issues like these, coupled with the advent of the computer, the growing doubts about the efficiency and effectiveness of state-controlled social provision of services, and the emerging findings of cognitive neuroscience call into question some of the fundamental building blocks of traditional education – schools, classrooms, teachers (as we understand the profession today), or even the curriculum, and even concepts like intelligence or ability.

While most people are more unsure about these issues than they were twenty years ago, there is no doubt that those who practise the art of education are likely to gain insights into human learning which will provide testable hypotheses for the scientists. The traffic between the cognitive neurosciences (including psychology) on the one hand and education on the other does not – and should not – all flow in one direction. The perceptions and experience of educators often identify issues which demand scientific investigation and explanation. Obvious examples might be the importance of early learning, self-esteem and motivation.

The following text has been designed for a well-informed general audience. It is intended to be accessible to non-specialists. It seeks to avoid exclusive language, professional disputes or territorial defence. But readers should be warned against the assumption that we all share a common language and a common conceptual framework. Terms like *plasticity* – central to brain science, unheard of in education – or *intelligence* – where the reverse almost seems to be the case! – demonstrate that different approaches lead to differing points of view. No matter. Those who map new lands always use the method of triangulation from several viewpoints. More dangerous are terms like *stimulation* – common to both disciplines, but not necessarily used to mean the same thing. It is necessary to proceed cautiously, reflect carefully, and see how far a collaborative report into learning and the brain can go at present.

Below are listed ten questions about human learning, which seem of fundamental importance and which a report such as this might be expected to illuminate. Each one focuses on issues connected with the promotion of successful learning – whether measured by academic or vocational attainment, social or workplace responsibility, or personal satisfaction.

1. What is the balance between nature and nurture in the promotion of successful learning?

Do our genes give us a life sentence? Or can we, for instance, learn to learn faster? Where should we strike the balance between the extremes of genetic determinism and the fiction that "anything is possible provided the child has a good home and school"?

2. How important are the early years to successful lifelong learning?

Those who believe that our early years experience is critically important in fostering positive attitudes, essential skills and a solid foundation for primary, secondary and adult education, are sometimes accused of subscribing to "the myth of early learning". What are the stages of development in the infant brain? How can we best encourage healthy growth?

3. How significant is the distinction between "natural development" and "cultural education"?

Children learn to walk and talk naturally. Unless severely disabled or wickedly maltreated, they all do it, and all do it at much the same age. Learning trigonometry or the tango is different: these do not happen naturally. There is no "normal age" for mastering such skills – and by no means does everybody do them. Although imitation is a key strategy in achieving both natural development and cultural education, they seem to be different processes. Are they? Is the brain programmed for natural development in some specific ways – and otherwise generally receptive (or not!) to the learning experiences in the category of "cultural education"?

4. If the distinction is significant, how can we best promote these two types of learning – "natural development" and "cultural education"?

If there is a "myth of the early years"[4] it arises from a failure to think through the implications of the distinction. "Natural development" seems to require no more than ordinarily decent conditions – "good enough" parents, a satisfactory home, sufficient food and drink. Love, stimulation, nourishment, exercise, conversation and a good environment probably provide all that an infant needs for healthy "natural development". "Cultural education" is another matter: for a start, you almost certainly need a teacher if you want to learn to read, dance, or drive a car. How much "cultural education" is appropriate to the early years, and how should we best provide it?

5. How far is the successful learning of specific attitudes, skills and knowledge age-related?

Developmental learning ("maturation") is obviously age-related. For example, puberty brings with it an increasing awareness of, interest in, and competence with the opposite sex. What about the attitudes, skills and knowledge that form the normal curriculum of "cultural education"? Determination, teamwork and

4. Bruer, J.T. (1999), *The Myth of the First Three Years: A New Understanding of Early Brain Development and Lifelong Learning*, Free Press, New York.

colours, for example, are learned in the nursery. What about reading, a second (or third) language, parenting, chairmanship, tolerance, wind-surfing, the piano, chess, calculus, first-aid, cooking, problem-solving, self-awareness, dance... and so on? Is the human brain especially receptive to some of these at certain ages? And if so, why and how?

6. Why is remedial education so difficult?

Perhaps this is the obverse of the previous question. Children who miss out on stages of development – perhaps through injury or maltreatment – find it difficult to catch up later. If you can't walk or talk by the age of ten, it will probably always be a struggle. Does the brain also become gradually less receptive to other forms of learning, particularly to the curriculum of "cultural education"?

7. What can be said about different "styles of learning"?

The key question seems to be, if people have different styles of learning, are they born with them or do they develop and learn them as they grow up? It is often said that some people prefer to learn through their eyes, others through their ears and a further group through touch and feeling. Though the truth is more likely to be that we all like a different mix of these modes of learning.[5] Other approaches to the idea of "styles of learning" speak of incremental learners, "end-first" learners, reflective or experimental learners, learners who favour one or more faculties (like language, number, music, for example), social or solitary learners, and so on. As yet, there is no coherent theory of learning styles.[6] What can the science of the brain teach us about this question?

8. What is intelligence?

IQ theory has dominated education for a century or more. In an extreme, simplistic and popular form it seems to claim that our intelligence is a single entity, fixed at birth, which provides a kind of glass ceiling limiting our potential for successful learning. This is surely an inadequate and inaccurate account of human

5. Moreover, common sense suggests that the learning style probably depends not only on the subject (the learner), but also on the object (the content), and on how the subject deals with the object.
6. The OECD Secretariat wishes to clearly dissociate itself from any interpretation in this publication which, based on the ideas of individualistic differences in the brain and of different learning styles, would try to link certain genes to IQ and hence, could have a racist connotation towards any group or groups of people within the human community. Such interpretations should be condemned, and the authors wish by no means to explore such beliefs, neither in this work, nor in the studies to be conducted in the area of brain research and learning sciences.

intelligence. And yet, some people do seem to be able to learn faster than others are – or perhaps they can learn *some things* faster than others? What is the difference between the brains of children we rashly label as "dim" or "bright"? Have we "multiple intelligences" or just one?

9. What is emotional intelligence?

If the limbic system of the brain is the seat of emotion (among other things) and the cerebral cortex is the reasoning faculty, what does it mean to talk about "emotional intelligence"? Does this refer to the natural maturation of our emotions or is it a question of their education or training? What are we to make of the paradox that while the theory of IQ seems implausible, it is apparently measurable – while "emotional intelligence" is not measurable, although palatable and seemingly plaisible.

10. How does motivation work?

What has science to say about our likes and dislikes? Why do people differ in what interests, excites, bores or repels them? What makes the difference in the brain between "merely wanting" and "really wanting" something? What happens when our motivation changes – or when someone else inspires us to aim for a new goal?

Important as they are, these questions may be too general. Reports such as this may well reveal valuable insights, but they can hardly be expected to offer a fully-articulated "new map" of learning. But they can be expected to be subversive of the *status quo*. The more we learn about the human brain, especially in the early years, the less comfortable we find ourselves with the traditional classroom model and imposed curriculum of formal education. This concern is particularly intense, for example, in seeking to evaluate the relative merits of the nurturing mother versus institutional care for babies, home schooling versus formal education for children, the natural interests of adolescents versus the rigour of a national curriculum. It seems doubtful whether current arrangements for the education of the young are best designed to enhance imagination and creativity,[7] self-reliance and self-esteem. For all ages, but especially for the young, there is a need to reconsider the importance of play, the role of stress (both challenge and threat), and the implications of human variety. This list could easily be extended.

7. During the Tokyo forum, Dr. Akito Arima [reviewing data from the Third International Mathematics and Science Study (TIMSS)] noted the need to inculcate a creative mindset among students from an early age (see the Tokyo report on the OECD website: *www.oecd.org/pdf/*M00022000/M00022657*.pdf*).

Part I

PREMISES

Chapter 1

The Education Context

I keep six honest serving men
(They taught me all I knew);
Their names are What and Why and When
And How and Where and Who.

Rudyard Kipling

"Education is a wreck: but you can find treasure in wrecks." This comment, made by a schoolboy some ten years ago, neatly captures the paradox of modern education: at one and the same time, precious and disappointing. The high hopes of those advanced societies who established in the 19th century universal, compulsory, free, elementary education for their peoples have not been fully realised. Instead, as so many young people tell us that they hate school,[1] they fail to learn the basics of literacy and numeracy to enable them to become employable; and they disrupt their classes, or play truant, or practise "intellectual truancy".

And yet, no one who has experienced the benefits of a good education doubts its value. Learning is a source of health, wealth and happiness. Education is a route to the good life. Learning pays – and learning empowers. Effective learning, starting at birth and continuing into old age, gives each individual the best hope of a successful life. The first priority of the new learning agenda is summarised in the phrase "lifelong learning for all". This phrase demonstrates how far ideas about learning and attitudes to education have changed in recent years. And they are still changing – not least, in the importance societies ascribe to them. During the latter half of the 20th century, human learning has risen from being a relatively minor concern for governments and their electorates to become a major issue world-wide – and now the first priority for many nations.

Wherever you look you can find evidence of this change. The media have an insatiable appetite for the subject of learning. The market in learning services is

1. See: *www.pisa.oecd.org* and OECD (2001), *Knowledge and Skills for Life – First Results from* PISA 2000, Tables 4.1 and 4.2, pp. 265-266.

growing apace. Governments wrestle with the challenges of introducing nursery education, improving schools and increasing access to higher education. Organisations and businesses of every kind are seeking to transform themselves into "learning organisations". Individuals create their own personal learning plans, and give reality to the rhetoric of "lifelong learning". Few would argue today that Disraeli was wrong when he said in 1874: "Upon the education of the people of this country the fate of this country depends". And yet the paradox remains to be resolved. What further reforms can help us do better than offer students "treasure in wrecks"? Or should we be contemplating revolutionary change in educational provision?

1.1. Why and who

Societies tend to develop through three phases: aristocratic, meritocratic, democratic. The first respects privilege, the second merit, the third humanity. Today there is little left of the privileged society of the past. Privilege is out of fashion. No one seriously argues that the best people are to be found in the "best" families – or that such an elite should be given the best education and offered the best jobs. Inasmuch as this still seems to happen it is because (unexpectedly, but not really surprisingly) the meritocratic society turns out also to favour the privileged.[2] But class, race, religion, sex and age are each in their different ways an inappropriate basis for educational discrimination in a meritocratic or democratic society.

The meritocratic principle gives power and influence to those who can demonstrate the highest ability. In a meritocracy, one of the principal functions of education is to sort people by "ability and aptitude". Aristocracies *know* who the best people are, and reward them accordingly. Meritocracies *search* for the best people, and then reward them generously. In either case, education and opportunities to learn beyond the elementary level are rationed and given only to the best.[3]

Of course, both aristocratic and meritocratic societies justify their selective educational systems by referring to three other relevant factors – employment needs, the range of intelligence and the presumption that able people learn best if they are segregated from less able people. The economy of the 19th century required large numbers of navies, factory hands and domestic servants – and relatively few managers, consultants or professors. Today, the reverse is true. As far as we can see, the 21st century will require ever more brain-workers and ever fewer

2. See: OECD (2001), *Knowledge and Skills for Life – First Results from* PISA 2000, Tables 6.1 (a, b, c), 6.2 and 6.3, pp. 283-287, Table 6.7, p. 291, Table 8.2, p. 309.
3. In the UK, for example, the 11+ examination and selective grammar schools were – and still, in part, are – the tools of meritocracy. Selection and restricted entry to post-compulsory schooling, or to higher education, serve the same purpose.

of us with nothing more to offer than our brawn. In developed nations only very few jobs today do not require literacy at least to the level needed to read the tabloid newspapers. That kind of mindless job is gradually disappearing. The workplace is slowly increasing its demands on the educational system, and on the lifelong personal learning of each individual.

Intelligence is – or should be – an embarrassing term for educators. Even if, of course, not everybody considers "intelligence" and "IQ" as equivalent, we talk as if we understood it, act as if IQ is measurable, classify our students with assurance – and yet the truth is that not much is clearly known about human intelligence. Popular and simplistic accounts of IQ theory teach us that our intelligence is a single entity, fixed throughout life, and (for most people) provides a sort of glass ceiling which prevents them from making progress in advanced learning. All three ideas are probably false. The work of Howard Gardner[4] has persuaded many of the idea of multiple intelligence. Daniel Goleman[5] has introduced the new concept of emotional intelligence (EI), which further complicates the picture. Whatever else it is, intelligence is undoubtedly complex.

Any number of individuals have demonstrated in their own lives and learning that the idea of a level of intelligence fixed and unchangeable over a lifetime is questionable, if not downright silly. Many people, having seemed dim at school, earned degrees later on from distance education institutions and/or later shone in the workplace. Likewise, some did well at school, only to struggle in adult life. While it obviously remains true at the gross level that some people learn faster than others, our speed of learning (which is probably a key element in the idea of intelligence) is deeply affected by other factors like confidence, motivation and the compatibility of the learning environment.

The idea that human intelligence is strictly limited or in short supply seems odd today. Forty years ago, very few went on to higher education in OECD countries. Today, more than 30% gain entry to universities and colleges. The "Robbins Report", published in 1963 in the UK, has been proved right: "If there is to be talk of a pool of ability, it must be of pool which surpasses the widow's cruse in the Old Testament,[6] in that when more is taken for higher education in one generation more will tend to be available in the next". As more and more people embark on, and succeed in, courses of advanced learning, the only thing to say with certainty about the limits of human intelligence (as measured by educational achievement) is that they are unknown and continue to exceed our expectations.

Such a view does not deny the probability that our genetic inheritance to an extent conditions our learning potential, or that the early formation of the brain in

4. Gardner, H. (1983), *Frames of Mind*, London.
5. Goleman, D. (1995), *Emotional Intelligence*, New York.
6. See I Kings 17, 10-16.

childhood plays a large part in influencing later learning, or that success tends to lead to success (and failure to more failure). What it does claim is that literally no one is incapable of further beneficial learning.

Although common sense would indicate it, it is not sure that able people learn best if they are segregated from less able people; but in any case, less able people seem to do much worse if they are segregated from able people.[7] For more than half a century a debate has raged between those who see the social advantages of comprehensive education and those who see the educational or social advantages (for the able) of selective education. They are both right, in some sense – though neither side finds it easy to do justice to the strength of the other's argument, maybe due to the fact that both sides are aiming at different goals, supported by different worldviews.

The issue of segregation is a critical one for the learning agenda. For, as we move forward from the meritocratic to the democratic society, the reasons for selective education tend to fall away. The democratic society seeks the fulfilment of all its members, not only those who are judged most able. It offers patterns of employment which demand and reward successful lifelong learning for all. It has a strong faith in, and high hopes for, the intelligence and learning potential of everyone. And it is disposed to reject segregation and selectivity – in spite of the perception (by some) of benefit (often for the same) in selective systems.

Whatever the outcomes of this debate, human groups tend to conform to the perceived norm. Segregated groups do this even more strongly than diverse groups. We might compare the behaviour of children in school or in the family setting. Or adults in the workplace or at home. The presence of the peer group – whether at school or at work – may lead to conformist behaviour. Our behaviour seems to be freer in a more diverse setting. We are better able to be ourselves – and fulfil our distinct potential – unconstrained by our peers.

Some assert that people who achieve exceptional things tend to experience in early childhood three critical conditioning factors: plenty of interaction with "warm, demanding adults",[8] an exploratory curriculum of learning that leaves the learner lots of room for experiment and initiative, and only limited access to peer groups which would have a negative impact in terms of learning. Of course, it is true that peer groups can be supportive and provide a positive challenge to the learner. But the possibility of an adverse effect is at least as strong (if not stronger) than that of a beneficial effect.

7. See OECD (2001), *Knowledge and Skills for Life – First Results from* PISA 2000, Table 2.4, p. 257 and Figure 8.4, p. 199.
8. See OECD (2001), *Knowledge and Skills for Life – First Results from* PISA 2000, Tables 6.5 and 6.6, pp. 289-290.

A democratic society, genuinely committed to the encouragement of lifelong learning for all its people, is faced with a great challenge in the system of education it inherits from the antecedent meritocratic society. Can a system designed to sort and reward the most able be reformed in such a way as to help everyone fulfil their (very diverse) potential? Or, if reform is impossible, is a kind of educational revolution on the agenda for learning?

Within a democratic society, while there may be agreement about the objective of encouraging and providing for the lifelong learning of all, there is likely still to be a good deal of disagreement about its purpose. Some see the strength of the economic argument. They believe that a "world-class workforce" will spearhead national prosperity and increased competitiveness in the global economy.[9] Others are more swayed by the argument from equity. They hope that a democratic learning society will help to remedy the inequalities inherited from the antecedent aristocratic and meritocratic models. A third group is intent on maximising human fulfilment. They recognise and accept the extraordinary range and diversity of outcome and achievement that is likely to result from investing in the lifelong learning of every individual. And, lurking amongst these three contrasting points of view is a fourth: those who consciously or unconsciously still find value in the idea of an elite, and seek to preserve some form of selectivity.

It is not possible to satisfy all these different demands at the same time. We shall have to choose. The arguments from economics and human fulfilment are compatible and persuasive. Lifelong learning for all may well reduce the inequalities of the aristocratic and meritocratic heritage, but new inequalities will probably replace them. True democrats reject elitism – while recognising wryly that they are often the beneficiaries of such a system. However, it will remain true that, if you do not know *why* learning matters, or *who* should enjoy it, you will struggle to find coherent answers to the many questions concerning the idea of lifelong learning for all.

1.2. What and when

What we should learn, and when we should learn it, turn out to be interrelated questions, just as *why learn*? and *who should learn*? proved to be in the preceding paragraphs. If the model of learning remains rooted in the primacy of "initial education" (with perhaps a limited role for "continuing education"), the curriculum

9. But at the same time, it has been shown [see OECD (2001), *Cities and Regions in the New Learning Economy*, Chapter 4, p. 37] that it is the secondary educational attainment of a given population that has the strongest impact on its economic performance. Moreover, this educational achievement is to a large extent related to the "level of inclusion" school systems provide [see OECD (2001), *Knowledge and Skills for Life – First Results from PISA 2000*]: in other words, equity in education, when volontarily and seriously practiced, seems to be not only compatible with economic performance, but also a crucial component of competitiveness.

will tend to be loaded in school and college with as much valuable material as possible, for fear that learners might miss their best opportunity for benefit. But if we really mean what we say when we talk about lifelong learning, it becomes possible to unload the curriculum of youth and spread the desirable curriculum over a lifetime.[10] Trigonometry, for example, or Japanese or the history and geography of Latin America, are all interesting subjects – but none of them is really essential to the initial curriculum of people living in Europe. What is?

In contrast to the existing model of a National Curriculum, which seems to try to cram in all desirable learning, we might consider a "minimum essential global curriculum". What might it consist of? Literacy (reading, writing, speaking and listening) in the mother tongue and in at least one other,[11] numeracy, cultural literacy (including the essentials of history, geography, science and technology, together with opportunities to develop skills in music, art, drama and sport), personal and social skills, values and ethics, learning how to learn (including, of course, elements of cognitive neuroscience: the nature of the brain, how the brain learns, etc.),... and what else? Such a "curriculum of essentials" would leave plenty of room and time for the faster learners to explore other subjects and range widely, while slower learners would at least have a good chance of learning what we all *must* know, understand or be capable of, to function effectively in life and work.

Traditionally, a curriculum consists of three elements: knowledge, skills and attitudes (KSA). And traditional educational curricula tend to value knowledge above skills, and skills above attitudes. Experience of life and work suggests a

10. "Sensitive periods" to acquire cognitive functions could prove a very useful tool to design this "desirable curriculum" in the future. See Dr. Hideaki Koizumi's remarks on the "reorganisation of educational systems based on neuronal plasticity" (and periodicity), in 4.5.3 below.

11. Two questions need to be raised here: Firstly: a person learning only one foreign language (which might not be sufficient, by the way) should probably learn English, as it is considered the "world language" today (the modern "lingua franca"); however, and especially if one considers that mastering a second (at least) foreign language becomes more and more necessary in terms of individual competitiveness, should English imperatively be privileged by this "curriculum of essentials" as the (chronologically) first foreign language learned by non-native English speakers? Secondly: should English native speakers be exempted from learning a foreign language, because their mother tongue is the "world language"? There is a temptation to respond positively, at least spontaneously. But it is not certain that to not include any foreign language in the curriculum would not have negative effects, on an individual as well as on a collective level. This is another story, though. But the issue will have to be addressed at some point, especially since it is likely that the acquisition of a foreign language (and especially an early acquisition of that kind) has a positive impact on brain-mapping (to say nothing of cultural open-mindedness), which would provide the individual with a comparative advantage (not only of technical nature); in that case, not having learnt any foreign language could possibly result in a comparative disadvantage.

different priority: ASK. Positive attitudes (such as responsibility, hopefulness, confidence and trust) are the key to the good life or a rewarding job. Skills (like communication, teamwork, organisation and problem-solving) are also essential. When so much of the world's store of knowledge is easily accessible in books or on the Internet, it becomes less important to be able to retrieve it from one's own brain.[12] The challenge is to create a learning society (not a "knowledge society"[13]) for the 21st century; a learning society requires an ASK curriculum.

On many subjects, the young brain learns faster than the old one; but adults are often better motivated to learn than children. On the whole, motivation is more important than youth to successful learning – though the combination is, of course, unbeatable. Perhaps we should consider an imposed "curriculum of essentials", as sketched above, coupled with the bold liberal principle of *trust the informed learner's demand* (TILD), once the essential curriculum is mastered. A nation that followed this precept would, of course, debate long and hard what exactly constitutes the "essential curriculum", and invest carefully in educational guidance.

What is clear is that "the best of the past" is no longer necessarily "the best for the future". In a society of little or no change, the wisdom of the elders and the experience of the past provide good guides for the young. But in an era of rapid and accelerating change this is no longer necessarily true. The young may be better placed than the old to judge what is essential, and what merely desirable, for them to learn. Somewhere between these two extreme views, an inter-generational dialogue might be highly desirable.

In adult life, the TILD principle should be our guide. Where we are in control of our own learning – in the home, in our leisure activities, in self-employment, or in retirement – this principle prevails. We learn what we choose to learn. The world of employment offers a different picture. Some short-sighted employers still do not recognise the value of learning for work. Others can see the value of vocational training and relevant skills, but doubt whether there is as yet a well-established business case for investing freely in the learning of the whole workforce. A few have recognised the truth of the claim that "learning pays" and are intent on moving towards the

12. This raises another question, this time about the content and the structure of knowledge to be acquired: this is the essential distinction between "know-what" (information, or "knowledge about facts") and "know-why" ("knowledge about principles and laws of motion in nature, in the human mind and in society"). It is a very complex debate, since "learning to learn" and acquiring any "know-why" cannot be achieved without a minimum of "know-what". If "it becomes less important to be able to retrieve [information] from one's own brain", the question about which basic information should be integrated in a "minimum, essential, global curriculum" remains widely open. [For more precise definitions of "know-what" and "know-why", and to go deeper into this debate, see OECD (2000), *Knowledge Management in the Learning Society*, especially pp. 14ff.]
13. It has been argued that every human society is a knowledge society, which seems to make a lot of sense; but not every human society is a learning society.

development of true learning organisations. In partnership with their employees they promote and encourage a liberal approach to learning and observe the TILD principle.

1.3. How and where

How do people learn best? And where do they best like to learn? Some people like to learn at home, others at work, others at college. The remarkable achievements of the "home-schoolers" could have revolutionary implications. There appear to be a multitude of learning styles, for example defined by medium (eye, ear or hand), or favoured intelligence-type, or gender, or preference for theory or practice, incremental or "end first" learning, and so on. We are nowhere near an adequate theory or practical analysis of learning styles as yet. What we do know is that successful learning is likely if the learner *a)* has high confidence and good self-esteem, *b)* is strongly motivated to learn and *c)* is able to learn in an environment characterised by "high challenge" coupled with "low threat".

Learning failure occurs when one (or more) of four impediments prevent the achievement of success. These impediments to learning are: *a)* lack of confidence and self-esteem (the feel good factor); *b)* weak motivation (not "really wanting" to learn); *c)* real (or perceived) inadequate potential ("It's too hard" or "I can't do it"); *d)* absence of opportunities to learn. Most educational debate addresses the last two, concerning itself with questions like the "ability range", IQ, aptitude – or access, equal opportunities, growth of provision. Important as these issues are, they are not necessarily the major impediments to learning in the developed world today. Concentrating on these almost to the exclusion of the first two was a sort of 20th century heresy. The primary problems for learners are confidence and motivation: this idea, widely shared among educators, could itself provide a fertile field for scientific research.

Confidence and self-esteem are necessary – but not sufficient – conditions for motivation (*really* wanting to learn). Any happy infant or self-assured adult demonstrates this truth. So the challenge for the learning agenda of the future is deceptively simple: to foster (or restore) the confidence and self-esteem that babies are born with. An environment characterised by a combination of "high challenge" and "low threat" does just this. Threats induce fear of failure; challenge encourages aspirations for success.

If "high challenge" coupled with "low threat" is ideal – and the reverse is certainly pernicious – the following matrix shows each of the possible combinations and the likely effect on the human learner (child or adult):

	High challenge	Low challenge
High threat	"anxious"	"dim"
Low threat	"bright"	"spoilt" or "indifferent"

Good education, effective training and successful learning take place in t lower left-hand corner of this figure. They develop and foster bright (alert, confi dent, self-assured, well-motivated and happy) children and adults, who are *mastery learners*. Those whose lives are lived, and whose learning is experienced, in the other three boxes become *dependency learners*, always relying on others for their standards, motivation and self-respect.[14]

Today we are on the threshold of a far deeper understanding of how people learn best – and how best to help them. Confidence and self-esteem (like milk and orange juice) are essential to the nourishment of the successful learner. These qualities are essential to effective motivation, but by themselves they are not enough. Well-motivated learners develop a burning passion for success: they understand the benefits of learning, they have dismissed any sense of personal inadequacy or inability, they have discovered for themselves good opportunities to learn, their highest priority is learning success. Undoubtedly, human motivation must be high on our learning agenda for the 21st century.

Kipling's "six honest serving men" provide a handy tool for revealing the outline of the learning agenda for the future. The central question is whether we can create a true learning society by means of the normal processes of gradual reform, so as to adapt our existing models and patterns of provision to meet the needs of the new century, or whether we need to think rather in terms of replacing them with something distinctly different. Discontinuous change is difficult to think about – until it happens. The dissolution of the monasteries, the development of air travel, and the discovery of the contraceptive pill can each be seen in retrospect as examples of discontinuous change, with revolutionary implications. Perhaps something similar is happening in education today.

There are several reasons for this – and some have been set out above. Three seem particularly important: the impending impact of the new "brain sciences" on our understanding of human learning, the computer – and the potential of Information and Communication Technology (ICT),[15] and the idea of "funding the learner" (rather than the teaching) in order to promote the effect of market forces on our social provision for learning, so as to raise the quality, improve the relevance and convenience, and lower the cost. The idea of funding education by means of "learning vouchers" for those intended to benefit, instead of grants to providers of education, seems worth considering.

14. But it is likely that mastery learners also rely on others to find the same things (notably motivation); however, this relative dependency is probably more positive: "reward" or "recognition" is likely to be sought for here. It might become useful, over time, to integrate this "reward" issue as a third element in the matrix above, besides "challenge" and "threat".
15. See: OECD (2000), *Learning to Bridge the Digital Divide*; OECD (2001), *Learning to Change: ICT in Schools*; OECD (2001), *E-Learning – The Partnership Challenge*.

But whether or not governments adopt the idea of "funding the learner", the learning revolution is underway and irreversible. ICT has already demonstrated its power to create a learning revolution by itself. The learning market is launched.[16] In the decades ahead we can expect to begin to unravel the complexities of the brain and understand at last the nature of memory and intelligence (for example) and, what exactly happens when learning occurs. When we do, we shall be able to refound our practice of education on a solid theory of learning. The outcome is more likely to be an example of discontinuous change than gradual adaptation of today's arrangements. Revolution, not reform.

16. OECD-CERI has the topic of "Trade in Educational Services" in its programme of work 2002-2004.

Chapter 2

How Cognitive Neuroscience Can Inform Education Policies and Practices

2.1. What cognitive neuroscience can tell...

"My brain? It's my second favourite organ."
Woody Allen

How do people learn? What happens in the brain when we acquire knowledge (names, dates, formulae) or skills (reading, dancing, drawing) or attitudes (self-reliance, responsibility, optimism)? Questions like these have interested humans for centuries. Today, scientists are beginning to understand how the young brain develops and the mature brain learns. Several disciplines contribute to this advance in knowledge. The most recently established, and probably the most important, is cognitive neuroscience.

As with most advances in science, the key is the development of new technology. Techniques[1] such as functional neuro-imaging, including both functional Magnetic Resonance Imaging (fMRI) and Positron Emission Tomography (PET), together with Transcranial Magnetic Stimulation (TMS) and Near Infrared Spectroscopy (NIRS), are enabling scientists to understand more clearly the workings of the brain and the nature of mind. In particular, they can begin to shed new light on old questions about human learning and suggest ways in which educational provision and the practice of teaching can better help young and adult learners.

It would be a mistake to promise or expect too much too soon. While some valuable insights and results are already available, it may take years before the findings of this new science can be safely and readily applied in education. But the subject will advance most successfully if the various disciplines which comprise "the sciences of learning" communicate and co-operate with one another. While it is already clear that it is beneficial for neuroscientists and educators to converse, seek to establish a common language, and challenge each other's

1. See 4.2. below and glossary for narrower definitions of the different technologies mentioned here.

hypotheses and assumptions, even greater benefit results from enlarging the debate to include both psychology and medicine. Cognitive psychology, in particular, has a pivotal role to play as a mediator between neuroscientists, on the one side, and educational practitioners and policy-makers, on the other.[2] But there is little doubt that, as a new "learning science" emerges in the years ahead, it will continue to draw on an even wider range of disciplines, including developmental and evolutionary psychology, anthropology and sociology.

Communication problems will exist among neuroscientists and educators. These two communities, generally, do not share a similar professional vocabulary; they apply different methods and logics; they explore different questions; they pursue different goals. They are perceived differently in the policy arena. Neuroscientists scientifically study the seat of learning itself: the brain. They carry with them the authority and aura of an arcane science. They are relatively few in number and employ expensive technology. By contrast, teachers of adolescents work in a complex social milieu in which their students may not share their goals. Their tools typically comprise chalk, talk, and textbooks. It is therefore necessary to be aware of the cultural differences between these two professions, and work to reduce misunderstandings and miscommunications and to promote understanding. Policy-makers can help to bridge the gap by promoting the professional sharing of resources, particularly the insights gained at the respective levels of analysis (*i.e.*, classroom learning and brain function), so that the findings of this emerging field may inform both our understanding of the brain-as-machine, and the brain-in-action (human learning). One of the difficulties to be faced is the need for a common language and shared vocabulary, among the diverse disciplines comprising "the learning sciences". Terms like *plasticity, intelligence* and *stimulation* (already identified as problematic in the introduction), exemplify the issue. To those, it would be easy to add a longer list, such as: *ability, attitude, control, development, emotion, imitation, skill, learning, memory, mind, nature* and *nurture*... The last two terms should remind us of the gap in understanding between the public perception of *nature* and *nurture* as two separate and autonomous domains and the scientific recognition of the mutuality of influence between them and the concept of the "experience-dependent development" of the "natural brain". The journey from genes to behaviour is long and arduous: somewhere near the centre lies the brain, both an expression of genetic material and the source of human behaviour.[3]

2. Moreover, one of the important revolutions coming from brain research in the nineties, confirmed by numerous presentations during the three OECD fora, is that studying the brain from "the outside", the goal of cognitive psychology, and observing the brain from "the inside", the goal of neuroscience, are in fact complementary. Cognitive psychology studies and discovers thinking and learning behaviours and helps generate hypotheses about the mechanisms that account for them, cognitive neuroscience directly studies and establishes (or confirms) what these mechanisms are.
3. See the Granada report on the OECD website: *www.oecd.org/pdf/*M00017000/M00017849*.pdf*

It is to be hoped that, as cognitive neuroscience continues to contribute to the emerging dialogue on science and learning, it will help to illuminate and resolve a number of awkward dichotomies, such as *nature* and *nurture*. *Plasticity* and *periodicity* are another pair of opposed ideas which need to be understood in a way that avoids a choice between them. Common sense and brain science confirm that our brains are *plastic* – they continue to develop, learn and change until advanced senility or death intervenes. The idea of lifelong learning makes sense. It is never too late to learn – provided the learner is well endowed with confidence, self-esteem and motivation. And yet there do seem to be *sensitive periods*, like "windows of opportunity" when the developing brain is particularly sensitive to certain stimuli and very ready to learn. An obvious example of this is the extraordinary speed with which young children acquire their first language. All children, apart from the most severely disabled, do this at much the same rate and time the world over, regardless of their later educational classification as slow or fast learners, of high or low intelligence, successes or failures. There may also be sensitive periods for "second language learning", the acquisition of social skills like team working, and even the critical choice between "mastery" and "dependency" learning. And yet, the brain is also persistently plastic.

Cognitive neuroscience will also help us understand the distinction between what is common to all human brains and our individual differences. For example, male and female brains appear to differ, but it is not at all clear what this implies. There are significant maturational differences between the young, adolescent and mature adult brains. As yet, cognitive neuroscience has little to tell us about individual differences. Moreover, at this early stage of the science, practitioners understandably find it in many ways easier to study disability (and outstanding ability) than the "normal brain". This need not be regretted, if it helps us to understand and support better those with conditions such as autism or Asperger's Syndrome. Actually, studying the brains of disabled (or outstanding) individuals is in fact one of the surest ways to gain insight into the workings of "normal" brains.

Scientists are understandably cautious, particularly in reporting conclusions in such a sensitive and exciting field as the human brain. It will help if there can be a general agreement to seek to distinguish between *a*) what is well-established (plasticity), *b*) what is probably so (sensitive periods), *c*) what is intelligent speculation (the implications of gender) and what is a popular misconception or oversimplification (the role of the "left and right hemispheres"). In any event, it seems likely that cognitive neuroscience will, in the years ahead, play a significant part in providing reliable answers to important questions about human learning, such as:

- What is the appropriate learning environment and learning agenda for very young children? In particular, is it advisable to provide an intensive early programme of training in numeracy and literacy (hot-housing)?

- What are the key sensitive periods in the development of the brain? What are the implications of these for an age-related curriculum of learning?
- Why do some people find literacy and numeracy so difficult to acquire? What can be done to prevent or remedy such conditions as dyslexia and dyscalculia?
- What are the limits of the human brain? Can anyone expect to match the achievement of people like Leibniz, Mozart or J.S. Mill, with appropriate teaching in the right environment?
- Why is unlearning so difficult? How can bad habits, incompetent skills, erroneous knowledge be corrected efficiently and effectively?
- What is the role of emotion in learning? How can we facilitate the limbic system (emotional) and cerebral cortex (cognitive) of the brain to co-operate when faced with a learning challenge?

2.2. ... the education policies

"Politik ist die Kunst des Möglichen."[4]

Bismarck

"Politics is not the art of the possible.
It consists of choosing between
the disastrous and the unpalatable."

J.K. Galbraith

Parents are the first educators of children. In many parts of the world, systematic education was – and still is – a function of religion. Throughout the world, in developed and developing nations, the state has assumed responsibility for providing, as far as possible, free and compulsory schooling for children, and access for more advanced learning for young and adult students. The first challenge for policy-makers is to balance and reconcile the roles of parents, religion and the state.

The second is to ensure that, as controlling and dominant partner in this trinity, its provision of learning opportunities in schools, colleges, universities and other educational and training institutions meets the needs of the learners, employers and the wider community at an acceptable cost. The cost-effectiveness of formal education and training is, perhaps, the central policy issue. And yet, it is fatally easy to assume that funding and standards are the only real questions.

4. "Politics is the art of the possible."

Since 1989 the world has been reflecting on the implications of the fall of communism. Democracy, the free society and market capitalism seem to have defeated the hopes of socialism. We have learned that free markets, controlled by a democratically governed legal framework, are both more efficient and more effective in giving people what they want and need than systems of central planning. Few dispute this in the case of the distribution of goods such as housing, food or clothes; or of services like entertainment, banking or hairdressing. Education, and in some parts of the world, health, are still seen as special services which cannot be safely entrusted to the marketplace. This question is the subject of lively debate in the case of health.

Although it could be useful to consider possible alternatives, it is assumed for the present that full state provision will continue to prevail in OECD countries, as it does largely throughout the rest of the world. Most states regulate and fund universal compulsory education and provide the schools, colleges and universities that are needed. Such states are much exercised by the question of cost-effectiveness – how to achieve the best results at the lowest cost. Most of the more detailed policy issues which recur in educational debates can be traced back to this fundamental question, *e.g.* class-size, extension of the period of (compulsory) educational entitlement, increased access to post-compulsory education, teacher-supply, entitlement to lifelong learning, qualifications and standards, inspection, etc. It seems likely that the scientific study of the brain and learning will make a significant contribution, not only to such detailed issues of policy, but also to the fundamental challenge of cost-effectiveness in education.

In recent years, there has been particular interest in the provision of good nursery education for all children, the revision and redefinition of the curriculum and the equipment of school classrooms with computers, to mention but three contemporary concerns. In each case, the underlying expectation was that there would be greater efficiency and effectiveness (a better cost-effectiveness ratio) and an improvement in standards greater than the increase in cost. Standards of educational achievement are notoriously difficult to measure; and reliable comparisons between nations or over time are hard to find. It is not easy to be sure that like is being compared to like. Nonetheless, there can be little doubt that there has been both an increase in costs and an improvement in standards over time, but it is not clear whether they are commensurate. What is clear is that the absence of any effective form of market competition in the world of education inevitably reduces its capacity to make gains in cost-effectiveness.

Policy-makers are also much exercised by the question of the purposes of education. One traditional view suggests that societies expect systems of education to do three things: to provide young and adult students with the necessary *skills*, to *socialise* them, and to *sort* them by ability and aptitude. These three functions are in tension with one another. If sorting is given priority, then the skills of slower learners tend to suffer, for example. If schools are to be responsible for the full range of 21st century skills and the socialisation needed for a complex modern society, then the curriculum is likely to become seriously overloaded. Sorting and socialisation are similarly uncomfortable bedfellows. Are schools expected to do too much?

Finally, it may be helpful to draw attention to the gap between the wants and needs of the learner and the concerns of policy-makers. The obsession with cost-effectiveness can easily distract policy-makers from a proper concern for, and understanding of, the complex nature of learning, the variety and sensitivity of learners, and the human brain.

Part II
COGNITIVE NEUROSCIENCE MEETS EDUCATION

Chapter 3

The Three Fora

The purpose of the OECD-CERI project on "Learning Sciences and Brain Research" has been to encourage collaborations between learning sciences and brain research on the one hand, and researchers and policy-makers on the other hand. Together, the potentials of and concerns about a possible "brain-based education"[1] in the future highlighted the need for dialogue between the different stakeholders. Once the conceptual basis of the project had been established, after one year of planning, initial discussions began with major research institutions, which led to the arrangement of three conferences or fora, devoted in turn to "early learning", "youth learning" and "adult learning" (with a strong focus on "learning in ageing"). Detailed reports on each of the three fora are available on the OECD website.[2]

This chapter is a chronological summary of the three fora; it is meant to introduce Chapter 4 and therewith make it more easily accessible. Hence, the scientific outcomes of the fora are briefly described hereafter, and laid out more fully in the following chapter.

3.1. Brain mechanisms and early learning: the New York forum

"The distance from the new-born baby to the five-year-old is a chasm;
from the five-year-old to me is just one step."

Tolstoy

The first forum was held in New York (USA) at the Sackler Institute on 16-17 June 2000. The central question which dominated the forum was the tension

1. On this front, Dr. Jan van Ravens compared, during the Granada forum, medicine and education: "An explicit effort is being made to reach 'evidence-based medicine': an overall eradication of intuition and popular belief in favour of a full application of medical knowledge in the daily practice. Education is ready for that kind of treatment: away from a curriculum based on tradition and political compromise, and towards a curriculum based on the evidence provided by learning sciences, on their turn based on the outcomes of brain research, as far as possible."
2. *www.oecd.org/oecd/pages/home/displaygeneral/*0,3380,EN-*document*-603-5-*no*-27-26268-0,FF.*html*

between the concepts of brain plasticity and periodicity, the idea that – while the brain continues to adapt throughout life – there are sensitive periods for learning specific things at certain ages. The forum received reports on a number of important issues related to early learning: language acquisition, early cognition, mechanisms of reading, mathematical thinking and emotional competence.

Research was presented on second-language learning, suggesting that the acquisition of grammar is partly time-constrained. "The earlier, the easier and faster." This finding suggests that second-language learning might be more effective in primary, rather than secondary education. However, the brain continues to be receptive to new semantic information throughout life.

Experience-expectant learning takes place when the brain encounters the relevant experience at the appropriate time, the sensitive period. Experience-dependent learning is sometimes constrained by age because sensitive periods may be present only during certain phases of development. Moreover, learning during a sensitive period may well require the appropriate environment. It appears, not unexpectedly, that brains respond better to complex environments than to those lacking stimulus or interest.

Children develop theories about the world extremely early and revise them in light of experience. The domains of early learning include linguistics, psychology, biology and physics – how language, people, animals, plants and objects work. Even at the moment of birth the child's brain is not a *tabula rasa*. Early education needs to take better account both of the distinctive mind and individual conceptualisation of young children and their preferred modes of learning, *e.g.* through play.[3]

It is probably in the realm of literacy that brain science can offer most to educators at present. Reading difficulties can arise from a number of causes, such as visual impairment, auditory weakness, or inappropriate strategies (cognitive dysfunction). None of these conditions places the child beyond help. When teachers and

3. According to Dr. Alison Gopnik (during the New York forum), infants come equipped to learn language. But they also learn about how people around them think, feel, and how this is related to their own thinking and feeling. Children learn everyday psychology. They also learn everyday physics (how objects move and how to interact with them), and everyday biology (how simple living things, plants, and animals work). They master these complex domains before any official schooling takes place.
Experts would like to see school practices build upon the knowledge children have gained in their earliest environments. For instance it might make sense to teach everyday psychology during early school. Or in the case of physics and biology, schools could start to teach children from their natural conceptions (and misconceptions) about reality in order to achieve a real understanding of the scientific concepts that describe it. Schools could capitalise more on play, spontaneous exploration, prediction, and feedback, which seem to be so potent in spontaneous home learning. Schools should be providing even the youngest children with the chance to be scientists and not just tell them about science (see the New York report on the OECD website, *www.oecd.org/pdf/M00019000/M00019809.pdf*).

scientists work together, there is real hope that we can provide early diagnosis and appropriate interventions to assist those at risk of various kinds of dyslexia.

Numeracy, like literacy, is a basic skill where cognitive neuroscience can come to the aid of education. Mathematical intelligence appears to be quite complex, involving several different parts of the brain, which are organised to work together by a control mechanism in the frontal cortex. Such a model suggests that there may be several different reasons (rooted in processing in different brain regions), why difficulties in numeracy arise.

The brain is the seat of emotion, as well as of reason. Indeed, our "emotional intelligence" (EQ) seems to be even more important to achievement and success than "cognitive intelligence" (IQ). The critical distinction between "mastery" and "dependency learning"[4] is a question more of (emotional) attitude than intellect. Successful learners seem to develop a form of self-control called "effortful control"[5] at an early age. In principle, this key skill can be instilled and encouraged, although it is significantly heritable.

The main scientific conclusions of the forum are set out more fully in Chapter 4. The forum reached five general conclusions related to the value and potential of trans-disciplinary[6] discussion, the distinction between new concepts and the scientific confirmation of old insights, the fundamental nature of the ideas of plasticity and periodicity, the relative importance of the early years for human learning, and the possibility of the emergence of a new *science of learning*.[7]

3.2. Brain mechanisms and youth learning: the Granada forum

"I would that there was no age between ten and three-and-twenty, or that youth would sleep out the rest. For there is nothing in the between but getting wenches with child, wronging the ancientry, stealing, fighting [...] would any but these boil'd brains of nineteen and two-and-twenty hunt this weather?"

Shakespeare

The second forum was held in Granada (Spain) on 1-3 February 2001. Two issues dominated the forum: the problem of translating the emerging findings of cognitive

4. See 1.3. above.
5. See 4.4.3. below.
6. Although these fora were organised before Hideaki Koizumi's concept of "trans-disciplinarity" (see Chapter 5) was adopted as a model for this project and for the dialogue it implies, the words "trans-disciplinarity" and "trans-disciplinary" will be preferred, throughout the text, to "interdisciplinarity" and "interdisciplinary".
7. See Chapter 5 below.

neuroscience in a form accessible to the world of education, and the idea of the adolescent brain as "work in progress".

It is easy to claim too much for cognitive neuroscience at this early stage in its development. Tentative findings derived from research on animals may be rashly generalised to produce confident statements about the human brain and learning. It is necessary to be cautious. It is probably best if neuroscientists and cognitive psychologists co-operate. Educators and policy-makers will gain most from, and contribute most to, a broader alliance of scientific research, including medicine. The science of learning is necessarily a trans-disciplinary one.

The adolescent brain may be viewed as "work in progress".[8] Brain imaging has revealed that both brain volume and myelination (a maturing process of neural connections) continue to grow throughout adolescence and, indeed, during the young adult period (20-30).

The forum also received reports on ADHD (attention deficit/hyperactivity disorder), the effects of some drugs on learning, brain damage and mathematical skills, implicit and explicit learning, mental imagery and mental stimulation, the acquisition of literacy, and other topics.[9] Once again, the main scientific conclusions of this forum are laid out more fully in Chapter 4.

There were also a number of more general conclusions. The first concerned the tension between the thirst of educators and policy-makers for new scientific insights and the caution of the scientists against making premature assumptions about human learning. The second came in an amusing debate about whether it constituted cheating to wear spectacles in a written examination. Obviously not. And yet, if one of the main functions of education is to sort people by ability and aptitude, how far is it appropriate to use these new scientific insights to help, for example, the slow reader or the poor calculator? Few disputed the view that giving people skills was far more important than sorting them. Third, the forum was much taken by the idea of the

8. Dr. José-Manuel Rodriguez-Ferrer, during the Granada forum, raised an hypothesis. He suggested an alternative way to understand some of the difficulties facing adolescents and young adults by relating psychological maturity with measures of the maturity of the prefrontal cortex, and not to attribute the typical social, behavioural, and psychological characteristics of adolescents to some possible hormonal substrates. His claim is based on brain imaging data that shows that the prefrontal cortex is slow to mature even among those in the 20 to 30 year age bracket (see Granada report, OECD website, *op. cit.*).
9. "How should the public be informed about the interpretation of these data? How should people think about the relative contributions of genetic and environmental factors? In particular, policy-makers should be informed about cognitive strategies and technological prostheses that can overcome learning deficits, whatever their basis. For example, if the findings related to attention deficit/hyperactivity disorder are borne out, they have clear policy implications for psychopharmacological interventions" (Dr. Jim Swanson during the Granada forum. See the Granada report on the OECD website, *op. cit.*).

teenage brain as an immature, if not incomplete, organ. This insight was matched by the undeniable claim that learning itself is learnable.

The forum concluded that the agenda had not done justice to some important issues, notably gender, culture,[10] self-esteem and peer influence. As in New York, participants discussed the emergence of a new science of learning, and confirmed the idea that at least five constituencies should be involved: cognitive neuroscience, psychology, health, education and policy-making. The model of interaction between these different constituencies should be more like a roundabout, or an ascending spiral,[11] than a one-way street. The time has come to move forward from communication to co-operation. For example, educators have much experience with problems that arise in classroom settings. These problems not only can specify research agendas for scientists, but successful solutions achieved "in the field" can provide strong hypotheses that can be formally tested "in the laboratory".

Finally, the forum revealed a number of challenges for scientists and educators. To aid public understanding of the complexity of the brain and learning, we urgently need a new and better model to help us grasp the interplay between nature and nurture,[12] plasticity and periodicity, potential and ability, etc. There was also a demand for a shift from "curriculum-led" educational systems to ones that are "pedagogy-led". The "how" of learning governs the "what".

Participants wondered whether it any longer made sense to link "skilling" with "sorting" in our schools. Perhaps the time had come to create two distinct services and professions. Some wondered whether the school-and-classroom model of

10. Subsequently, these two issues were addressed in Tokyo (see below and the Tokyo report on the OECD website, *op. cit.*).
11. See Koizumi's trans-disciplinarity model in Chapter 5.
12. Dr. Antonio Marín discussed during the Granada forum (see Granada report, OECD website, *op.cit.*) the nature-nurture controversy and its relationship to the heritability of intelligence, mentioning early work by Dr. Francis Galton that studied the frequency of eminent individuals among relatives of outstanding men. He also touched upon the corrupting effects of the eugenics movement on the question of the heritability of intelligence. He pointed out not only the role of genetic, but also environmental factors: intellectual performance is a result of many years of training that involves influences of parents, teachers and other people (notably peer-influenced learning): "The genes determining the measurable trait cannot be identified, and it is not possible to make any specific inferences regarding their number, mode of inheritance or mode of action." But he believes that we can expect, over time, that the genetic variability of biological factors influencing learning ability and other aspects of human behaviour may be as extensive as the influence of genetic variability on human health. Progress on this front can already be seen in experimental analyses of animals. On the other hand, in humans, the Genome Project is finding much genetic variance. Dr. Marin also ended his talk with a caution against falling into a naive biological determinism, in which individuals are viewed as limited by their genes.

education derived from the outdated 19th century factory model of industrial production, was sustainable as the central strategy for rearing the young in the 21st century societies.

3.3. Brain mechanisms and learning in ageing: the Tokyo forum

"Si jeunesse savait, si vieillesse pouvait…"

Henri Estienne

The third forum was held in Tokyo (Japan) on 26-27 April 2001. Although the forum had been created to consider questions relating to brain mechanisms in the adult population as a whole, the central issue turned out to be the nature of the ageing brain and the challenge of extending and enhancing cognitive functioning into old age. Up to now, little research has been directed at the learning needs of *normally ageing* adults: for example, how to retrain teachers,[13] or adults in general, in order to use new technologies.[14] The need for continuous education in adulthood is clear especially among professionals, when one considers the data on the limited "shelf-life" of new findings in the sciences (findings tend not to be cited after about five years[15]). Adults in the future will not only have to learn more, they will have to unlearn more.[16] While research in this area is still in its infancy and not enough is yet understood about the normal processes of ageing, it appears that there is real hope for early diagnosis and appropriate interventions to defer the onset or delay the acceleration of neuro-degenerative diseases in old age. Lifelong learning appears to offer a particularly effective strategy for combating senility and conditions such as Alzheimer's disease.[17]

The forum received reports on a range of disorders associated with ageing, strategies for arresting decline and enhancing capacity in those affected, the evidence that brain-plasticity endures throughout life, the relationship between physical fitness and mental health, questions of memory and attention, issues of

13. Dr. Eric Hamilton during the Tokyo forum (see the Tokyo report on the OECD website, *op. cit.*).
14. Mag. Wolfgang Schinagl during the Tokyo forum (see the Tokyo report on the OECD website, *op. cit.*).
15. Dr. Kenneth Whang during the Tokyo forum (see the Tokyo report on the OECD website, *op. cit.*).
16. During the Tokyo forum, the presentations by Dr. Bruce McCandliss and Andrea Volfova, showed that plasticity cuts both ways and can interfere with new learning in some cases (see the Tokyo report on the OECD website, *op. cit.*).
17. But we still lack a reliable method of detecting Alzheimer's Disease in its preclinical stages, as reported by Drs. Akihiko Takashima and Raja Parasuraman during the Tokyo forum (see Tokyo report, OECD website, *op. cit.*).

methodology, culture,[18] gender,[19] and policy in relation to research and health. The main scientific conclusions of this forum, like the others, are set out more fully in Chapter 4.

Although the range of debate was very wide-reaching ("from genes to skills") it was felt that insufficient attention had been given to questions of disposition – the development of attitudes and values, and the emotional component of human experience and learning. There was a strong desire to move forward from conversation to co-operation and promote a new trans-disciplinary research programme. There was a good deal of caution against expecting too much too soon, but at the same time there was also real hope of substantial longer-term gains.

It was recognised that the emergence of a new learning science was already creating policy issues for the advancement of science, the reform of education and the development of health care. But these could not be effectively considered in isolation. A holistic approach was required – and this would not be easy to achieve.

Alongside the exciting hope of better diagnosis, postponement and remediation of senility, the forum identified five opportunities for future trans-disciplinary work: reading, mathematics, gender, the measurement of ability, and the development of a new teaching profession. Underlying all of this was the perception that the emotions may provide the key to understanding how we may best raise our young, and care for the aged in the 21st century.

18. The influence of culture may be seen at both macro and micro levels. At the macro level, the question of culture can help to orient the research agenda and ask if cognitive neuroscience should be seeking universal processes or if studies of the brain and of learning are cultural through and through, and thus more situated (Dr. Shinobu Kitayama during the Tokyo forum; see Tokyo report, OECD website, *op. cit.*). At the more specific level, it can be shown that the different orthographies of different languages can impact the appearance of reading difficulties, particularly dyslexia, with important consequences for how phenomena such as dyslexia should be studied and explained.
19. The results associated with gender are equivocal, but data supporting the notion of "male" and "female" brains (with some suggestions about learning) are emerging, as suggested by Dr. Yasumasa Arai during the Tokyo forum (see Tokyo report, OECD website, *op. cit.*).

Chapter 4

Learning Seen from a Neuroscientific Approach

Exciting discoveries in cognitive neuroscience and continued developments in cognitive psychology are beginning to offer interesting ways of thinking about how the brain learns. Historically, both theory and method have separated these disciplines. However, with the development of new brain imaging technologies, a new combined science has emerged: cognitive neuroscience. Cognitive neuroscientists have been paying increasing attention to education as an area of application of cognitive neuroscience knowledge as well as a source of important research issues.

In this chapter, cognitive neuroscience research initially presented during the three international OECD fora is summarised. Though many individual topics were discussed, the main ones considered to have high potential application in addition to potentially contributing meaningful information in curricular design, teaching practices, and learning styles will be reported here. These topics include: literacy and numeracy, the role of emotion in learning and lifelong learning. But before going into the substance itself, it seems necessary to present briefly, along with some basic principles of the brain's architecture, the research tools (technology) and the methodologies used today in cognitive neuroscience. At the end of this chapter, "neuromythologies" based on popular misconceptions and/or misunderstandings of science will be reviewed.

4.1. Principles of brain organisation and neural information processing

4.1.1. *Neurons, mental states, knowledge and learning*

The basic component of information processing in the brain is the neuron, a cell capable of accumulating and transmitting electrical activity. There are approximately 100 billion neurons in a human brain, and each may be connected to thousands of others, enabling information signals to flow massively and in many directions at once.

At any moment, a very large number of neurons are active simultaneously, and each such so-called "pattern of activity" corresponds to a particular mental state. As electricity flows through the connections among neurons (called synapses) another set of neurons is activated and the brain shifts to another mental state. Contrary to computer bits which are either on or off, a neuron's activation level is a continuous variable, enabling incredibly subtle variations and shading in mental states (Figure 1).

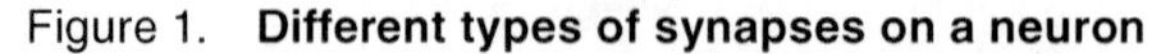

Figure 1. **Different types of synapses on a neuron**

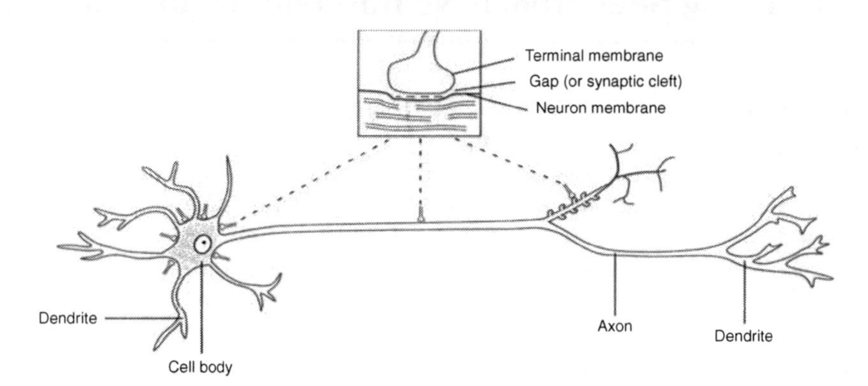

Note: The synapse includes the neuron membrane, the membrane of the terminal and the gap in between these two structures.
Source: Jean-Pierre Souteyrand for the OECD.

If mental states are produced by patterns of neural activity, then "knowledge", defined as whatever drives cognitive flow from one mental state to another, must be encoded in the neural connections. That means learning is achieved either through the growth of new synapses, or the strengthening or weakening of existing ones. Actually, there is good evidence for both mechanisms, with special emphasis on the first in young brains, and on the second in mature brains. It is perhaps worth noting that entering any new long term knowledge in a brain requires a modification of its anatomy.

4.1.2. *Functional organisation*

Different parts of the brain carry out different information processing tasks. This principle of functional localisation holds true at almost every level of brain organisation. The brain is a set of structures that sits on top of the spinal cord. The lower structures are devoted to co-ordinating basic bodily functions (*e.g.*, breathing, digestion, voluntary movement), expressing basic drives (*e.g.*, hunger, sexual arousal) and processing primary emotions (*e.g.*, fear). The higher structures, which evolved later and on top of the lower ones, are more developed in humans than in any other animal. The most recently evolved part, the neocortex, is a thin sheet of neurons that coats the convoluted surface of the brain. It is where thinking is done, and where three-fourth of the neurons in a human brain reside.

The neocortex is divided into two hemispheres, left and right. In between, a neural band of fibers called the corpus callosum acts as bridge, enabling the hemispheres to exchange information. Each hemisphere is further divided into lobes which are specialised for different tasks: The frontal lobe (front) is concerned with planning and action. The temporal lobe (side) is concerned with audition, memory, and object recognition. The parietal lobe (top) is concerned with sensation and spatial processing. The occipital lobe (back) is concerned with vision (Figure 2). These are gross characterisations, of course, as each lobe is further subdivided into interlocking networks of neurons specialised for very specific information processing. Any complex skill, like addition, or word recognition, depends on the co-ordinated action of several of these specialised neural networks localised in different parts of the brain. Any damage to one of these networks or to the connections among them will disrupt the skill they underlie, and to each possible anomaly corresponds a specific deficit.

Figure 2. **Major subdivisions of the cerebral cortex**

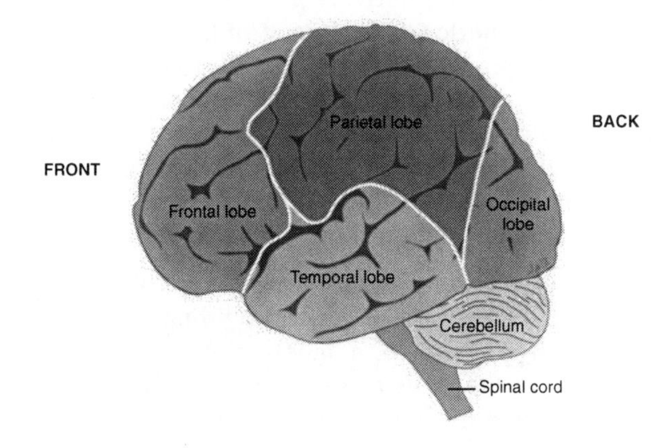

Source: Jean-Pierre Souteyrand for the OECD.

Finally, it should be noted that no two brains are alike. Significant individual differences are found in overall size, but also, more importantly, in the number of neurons assigned to carry particular functions, or even in the organisation and localisation of functional modules. Because most neurons are functionally interchangeable, the same neuron can be assigned to one task, and later reassigned to another, which means that nature, nurture, and learning cannot help but make every brain unique, and a work in progress throughout life.

4.2. Research tools, methodologies and educational implications: the impact of brain imaging

Neuroscientific research techniques vary and can include invasive procedures, including neurosurgery. However, the now most well-known and used tools are non-invasive brain-imaging technologies. Brain-imaging tools can be divided into two general categories, those that provide high-resolution spatial information and those that provide high-resolution temporal information about brain activity. Among those tools that provide high-resolution spatial information about brain activity, the best known are Positron Emission Tomography (PET) and functional Magnetic Resonance Imaging (fMRI). PET techniques, using radioisotopes, detect brain activity by monitoring changes in oxygen utilisation, glucose utilisation, and cerebral blood flow changes. FMRI, with the use of radio frequencies and magnets, identifies changes in the concentration of deoxygenated haemoglobin. Both techniques require subjects to remain motionless for accurate imaging.

Because PET and fMRI provide spatial resolution in the millimetre range, but temporal resolution only in seconds, these techniques are useful for measuring changes in brain activity during relatively prolonged cognitive activity. Another technique, Transcranial Magnetic Stimulation (TMS), is used to create a temporary disruption of brain function (a few seconds) in order to help locate brain activity in a circumscribed region of the brain. Nonetheless, processes such as performing mathematical calculations or reading involve many processes that occur over the course of a few hundred milliseconds. For that reason, PET and fMRI are able to localise brain regions involved in reading or mathematical activity, but cannot illuminate the dynamic interactions among mental processes during these activities.

Another set of tools provides accurate temporal resolution in the millisecond range, but their spatial resolution is coarse, providing data only in centimetres. These techniques measure electric or magnetic fields at the scalp surface during mental activity. Among these tools are electroencephalography (EEG), event-related potentials (ERP), and magnetoencephalography (MEG). EEG and ERP use electrodes placed on particular areas on the scalp. Because of their ease of use, these techniques are often used successfully with children. MEG uses super-conducting quantum interference devices (SQUIDs) at liquid helium temperature. Using these tools, accurate measures in the milliseconds of changes in brain activity during cognitive tasks can be measured.

A new method for non-invasive brain function imaging is optical topography (OT), which was developed by using near infrared spectroscopy (NIRS). Unlike conventional methodologies, it can be used for behavioural studies because the flexible optical fibres allow a subject to move, and a light and compact system can be built. This method can be applied to infants as well as adults. The observation of early development on a monthly time-scale will provide information about the architecture of the neuronal processing system in the brain. Optical topography may bring important implications for learning and education.[1]

Effective research in cognitive neuroscience requires a combination of these techniques in order to provide information on both spatial location and temporal changes in brain activity associated with learning. In making the link with learning processes, it is important for the neuroscientist to have fine-grained elementary cognitive operations and analyses in order to make powerful use of brain-imaging tools. Among those disciplines associated with learning, such fine range and analyses are most typically available from studies in cognitive science or cognitive psychology, and, to date, typically in studies of visual processing, memory, language, reading, mathematics and problem-solving.

Other research options available to neuroscientists include examining brains during autopsy (for example, to measure synaptic density) and in some rare cases, working with certain medical populations, such as those suffering from epilepsy (to learn about brain processes from people who have suffered brain damage or brain lesions due to disease or injury[2]). Some neuroscientists study children suffering from fetal alcohol syndrome or Fragile X syndrome, and others study the

1. Koizumi, H., *et al.* (1999), "Higher-order brain function analysis by trans-cranial dynamic near-infrared spectroscopy imaging", *Journal Biomed. Opt.*, Vol. 4, front cover and pp. 403-413.
2. Dr. Luis Fuentes referred during the Granada forum (see Granada report, OECD website, *op.cit.*) to the problem of linking neural theories of mind to the cognitive functions of the brain, and to an approach to the study of learning which values task decomposition, and advocates the study of simple cognitive tasks that are believed to involve the orchestration of elementary operations localised in the brain. Over the past 40 years, these methods have been used to study how people read, write, visualise, recognise objects, and so forth. The study of patients with brain lesions has lent support to the idea that different parts of the brain perform different computations. For example, visual agnosic patients have difficulty recognising intact faces while other visual ability is unimpaired. Some patients recognise exemplars of a specific semantic category while not recognising others. Dr. Fuentes believes that this accumulating evidence demands that complex cognitive systems be decoupled into elementary operations: "Firstly primates' studies, and then studies with patients with lesions in the parietal cortex, demonstrate that awareness depends on the orchestration of three different operations involved in shifting attention: disengaging, movement, and engaging of attention, performed by the posterior parietal lobe, the superior colliculus and the pulvinar nucleus of the thalamus, respectively. When patients present a lesion of any of these parts of the brain, they neglect stimuli appearing in the contralesional side of the lesion. In other words, they loose awareness of information presented to that part of the visual space despite the [fact that they] do not have any other vision problem."

cognitive decay prevalent during the onset of Alzheimer's disease or senile depression. Still others study the brains of primates or of other animals, such as rats or mice, in order to better understand how human mammalian brains function. In the past, without brain-imaging techniques available, it has been difficult to collect direct neuroscientific evidence of learning in the general, healthy human population.

A further limitation is presented by the fact that no single set of well-understood developmental learning tasks has been applied to normal human populations across the life-span. Much work has been carried out in regards to early childhood learning, but less regarding adolescent learning and even less again regarding adult learning. Without a baseline of normal cognitive development, it is difficult to understand any pathological occurrences in learning.

Understanding both the power and limitations of brain-imaging technology and the necessity of conducting rigorous cognitive protocols is the first step in trying to understand how cognitive neuroscience can guide education eventually in the formation of brain-based curricula. Recent findings are beginning to show that eventually education will emerge at the crossroads of cognitive neuroscience and cognitive psychology along with sophisticated and well-defined pedagogical analysis. In the future, education will be trans-disciplinary, with an intersection of different fields merging to produce a new generation of researchers and educational specialists adept at asking educationally significant questions at the right grain size.

Current research methods in cognitive neuroscience necessarily limit the types of questions that are addressed. For example, questions such as "How do individuals learn to recognise written words?" are more tractable than "How do individuals compare the themes of different stories?". This is because the first question leads to studies where the stimuli and responses can be easily controlled and contrasted with another task. As such, it becomes understandable in reference to known cognitive models. The second question involves too many factors that cannot be successfully separated during experimental testing. For this reason, the type of educational tasks favoured by society will remain more complex than the ones that might suit cognitive neuroscience.[3]

Researchers also stress the methodological necessity of testing for learning not only immediately after some educational intervention (which is typical of current practice), but also at certain intervals thereafter, especially in the case of age-related comparisons.[4] These longitudinal studies take the research projects out of

3. Dr. Barry McGaw, during the Tokyo forum (see Tokyo report on the OECD website, *op. cit.*)
4. Drs. Raja Parasuraman and Art Kramer during the Tokyo forum (see Tokyo report on the OECD website, *op. cit.*).

the laboratory and into real-life situations, which places limits on when the results can be interpreted and available for education use.

When attempting to understand and analyse scientific data, it is important to retain critical standards when judging claims about cognitive neuroscience and its educational implications. Some points to consider:

- the original study and its primary purpose;
- if the study is a single study or a series of studies;
- if the study involved a learning outcome;
- the population used.[5]

The importance of developing an informed critical community for the progress of science (that comes to consensus, over time, on the evidentiary and inferential basis of purported scientific claims) has recently been re-emphasised.[6] The development of such a community (composed of educators, cognitive psychologists, cognitive neuroscientists, and policy-makers, etc.) around the emerging sciences of learning is crucial. In order for that community to develop, an appropriately critical judgement in matters of "brain-based" claims about learning and teaching is necessary. Integrated into this community, education policy-makers will more successfully enter into appropriate brain-based curricula if there is a recognition of the following:

a) the popularity of a neuroscientific claim does not necessarily imply its validity;

b) the methodology and technology of cognitive neuroscience is still a work in progress;

c) learning is not completely under conscious or volitional control;

d) the brain undergoes natural developmental changes over the life-span;

e) much cognitive neuroscience research has been directed at understanding or addressing brain-related pathologies or diseases;

f) a satisfactory science of learning considers emotional and social factors in addition to cognitive ones; and

g) although a science of learning and brain-based education are just beginning, important gains are already being made.

There are ample data at the psychological level (drawn primarily from well-designed studies in cognitive psychology) from which to draw lessons for learning

5. Whether using human primates or non-human primates, questioning the representativeness of the sample, and asking to what population the claims do apply, is of utmost importance.
6. In a US National Research Council report on Scientific Inquiry in Education (*www.nap.edu*).

and teaching.[7] Data from cognitive neuroscience can help by refining hypotheses, disambiguating claims, and suggesting directions for research.[8] In other words, a major contribution of cognitive neuroscience to an emerging science of learning may be to imbue the discipline with a scientific scepticism toward unfettered claims and unexamined advocacy about how to improve teaching and learning.

But scepticism toward some current claims about the neuroscientific basis for learning should not breed cynicism about the potential benefits of cognitive neuroscience for education. Indeed, the emerging data about brain plasticity are encouraging. The evidence for claims about learning is unlikely to come from neuroscientific studies alone, however. In the future, improved brain imaging technologies and more sophisticated learning protocols may allow us to further illuminate this question.

4.3. Literacy and numeracy

4.3.1. *Language learning*

In the New York forum, literacy included both language learning and reading, as these are areas in which cognitive neuroscience can offer both insight and amelioration for problems such as second language learning and dyslexia. Dr. Helen Neville noted that second language learning involves both comprehension and production and accordingly, the mastery of different processes is necessary. Two of these processes,[9] grammar processing and semantic processing, rely on different neural systems within the brain. Grammar processing recruits more frontal regions of the left hemisphere, whereas semantic processing (such as vocabulary learning) activates the posterior lateral regions of

7. As Dr. Bruer pointed out during the Granada forum (see the Granada report on the OECD website, *op. cit.*).
8. Talking about the debate on the existence and character of implicit vs. explicit learning and its relationship to instruction, during the Granada forum (see Granada report, OECD website, *op. cit.*), Dr. Pio Tudela illustrated how cognitive neuroscience research could be used to help clarify and explicate debate among cognitive psychologists about the existence and characteristics of dissociable human learning systems. When a person learns about the environment without intending to do so and learns about it in such a way that the resulting knowledge is difficult to express, this process is often referred to as "implicit learning". By contrast, learning in which intentional attention is paid to the encoding of knowledge and in which retrieval is more conscious is called "explicit learning". Dr. Tudela showed that the results of neuropsychological studies (research with amnesic patients, Parkinson disease and Huntington's disease patients) and experiments using imaging techniques indicate different neural circuitry supporting implicit as opposed to explicit learning.
9. Two main processes are mentioned here; however, language involves other processes, referred to by Dr. Neville, among which understanding context and intent, prosody, and phonology (see the New York report on the OECD website, *op. cit.*).

both the left and right hemispheres. Language, in general, is not processed by a single region of the brain but by different neural systems located throughout the brain. This becomes interesting for educational application: the identification of the brain regions recruited for processing language gives insight on the impact of delaying exposure to a second language on these various subsystems.[10]

Research has shown that the later grammar is learned, the more active the brain becomes (more brain activation very often means that the brain finds that particular task more difficult to process: for example, expert readers will show less brain activation than novice readers in a word-recognition task). Instead of processing grammatical information only with the left hemisphere, late learners process the same information with both hemispheres. This change in brain activation indicates that delaying exposure to language leads the brain to use a different strategy when processing grammar. Confirmatory studies have additionally shown that subjects who had this bilateral activation in the brain, had significantly more difficulty in using grammar correctly. In other words, bilateral brain activation, in this case, probably indicates a greater difficulty in learning (which is confirmed by common experience).

Concerning second language learning, the earlier the child is exposed to the language, the easier and faster the grammar is mastered. Semantic learning, however, can and does continue throughout the life-span and is not constrained in time. Research on grammar learning is an example of both a sensitive period of learning as well as an experience-dependent one. Efficiency and mastery are not necessarily lost, but are just more difficult for the late learner, because relevant experience has not been received within a biologically defined time frame.

One clear educational policy consequence from this research area is that learning a second language (whose grammar markedly differs from one's own native language – for instance learning English for a native French speaker) after 13 years of age is extremely likely to result in poor mastery of the grammar of this language. This result is at odds with the education practices in numerous countries where second language learning starts at approximately 13 years of age. Another strong educational policy consequence of this is that if it is possible to identify which subsystem(s) of the brain is (are) subject to sensitive period constraints and which is (are) not, the development and implementation of sensible educational and rehabilitation programmes could become a goal for education policy-makers. It is one thing to know, globally, that learning a language later is usually more difficult, but another to establish this in such a way that public education policy decisions can be based on it.

10. Delay can also occur with first language learning in extreme cases, when children are not spoken to directly or when their verbal output is ignored.

Any public policy decisions concerning second language learning and any remediation (for instance to improve language learning for late language learners) will have to take into account how the brain processes language in order to insure effectiveness.

As is often the case in science, established assumptions are sometimes challenged and this is the case concerning second language learning among adults. It is known that native speakers of Japanese, when attempting to distinguish between the English sounds /r/ and /l/ (for example, in the words "Load" and "Road"), have considerable difficulty. The fact that these difficulties persist even after many years of experience in an English-speaking country provides support for the idea that there must be a "sensitive period" for the acquisition of phonetic contrasts.[11] The problem with this idea, according to Dr. Bruce McCandliss, is that this could generate the erroneous inference that learning deficits become permanent outside the sensitive period. To demonstrate that this was not the case, and that new learning in this field could occur in adults, research was conducted, in which the speech inputs of /r/ and /l/ were modified to such an extent that Japanese natives were able to perceive them as distinct inputs. With short-term training, subjects were able to transfer this ability when listening to unmodified speech. Complimentary neuroimaging results provided initial evidence that such training impacts the same general cortical regions implicated in native language speech perception.

4.3.2. *Reading skills*

When children arrive at school anywhere between the ages of 4 and 7, they are already experts in visual object recognition and at converting sound inputs into language representations. They have specialised neural connections for these skills, which are genetically programmed. Additionally, they have a full command of the syntax and comprehension of sentences as well as complex sentence contexts. With the shift from struggling to sound out words to automatically recognising words, or learning to read, different brain mechanisms are activated.

This is an important insight for education as this could hold important implications for interventions directed towards young and adult readers who have trouble with word recognition. Generally, when school-aged children do not successfully read they are thought to have a disorder called dyslexia. At least one

11. As was mentioned during the New York forum by Dr. Neville in support of a "sensitive period" for phonology (or the process by which the sounds of language are perceived, produced, and combined): "When you hear someone speaking your language with a non-native accent, you can be sure that he [or she] learned it after the age of 12" (see the New York report on the OECD website, *op. cit.*).

brain region appears to be critical in distinguishing the dyslexic reader. This region, the left superior temporal gyrus, attends to the sound structure of words at the level of phonemes. Researchers have found that 10-year-old children with dyslexia fail to activate this brain region normally during tasks associated with reading and phonological skills. Instead, these readers show a greater than normal amount of activity in the frontal region of their brain, which may reflect their attempts to compensate for their deficit. With on-going studies in dyslexia, neuroscientists and educators alike start to understand why children considered as normally intelligent still cannot read, or at least experience important difficulties when learning to read.

A first impulse upon discovering that a learning difficulty is due to a "brain problem" is to consider it beyond remediation by purely educational means. However, one can also turn this around and consider that when the breakdown of a skill into its separate information processing steps and functional modules is sufficiently understood, thanks to the tools of cognitive neuroscience, that is when efficient remediation programmes can be devised. This is precisely what Drs. Bruce McCandliss and Isabelle Beck did in the case of dyslexia, building on the intact components of reading skills in dyslexic children to come up with a new method for teaching word pronounciation. And of course, such deep understanding of how a skill is decomposed into separate cognitive processes may also help design better methods for teaching unimpaired children.

Using their "Word Building Method", Dr. McCandliss and Dr. Beck showed that dyslexic children are capable of learning to read. Helping children to generalise from their reading experience enables them to transfer what they had learned about specific words to new vocabulary words. These skills involve alphabetical decoding and word building and enable reading impaired children to progressively pronounce a larger and larger amount of words. This method teaches them that with a small set of letters, a large number of words can be made. As many of school aged children have difficulty in reading, attending to this problem allows this substantial portion of learners to engage in the most fundamental linguistic exchange and lessens their potential marginalisation from society. Others, most notably Drs. Paula Tallal and Michael Merzenich, have reported similar findings with a different technique. Although these results are somewhat controversial, their method does appear to help at least some children. The key point, however, is not whether one particular available method works better than others. Rather, we note that the theoretical and methodological machinery exists to attack the problem, and progress is clearly being made. Many, like for instance Dr. Emile Servan-Schreiber, predict that the study and treatment of dyslexia will be one of the major "success stories" of cognitive neuroscience in the relatively near future.

4.3.3. *Mathematical skills*

Mathematical thought, which includes how humans think about and manipulate numbers, is almost always difficult for children entering school. However, according to Dr. Stanislas Dehaene, infants do have an innate number sense and this constitutes an elementary number theory. The part of the brain responsible for this ability, the intraparietal sulci, is specialised for representing numbers as a quantity and enables infants to understand the difference between "a lot" and "a few". Learning mathematics in school encourages children to exceed their innate approximation skills and to distinguish between different numbers and perform arithmetic operations and manipulations.

Recent research in mathematics learning and cognitive neuroscience has shown that the brain recruits different regions to accomplish different tasks in mathematics.[12] The Triple Code Model[13] posits that for three basic number manipulations, there are three different brain areas that are recruited. When seeing a visual digit, *e.g.*, "3", the fusiform gyrus is active. When hearing or reading the number as a word, "three", the perisylvian area is active and when understanding a number as a quantity, *i.e.*, "3 is bigger than 1", the interparietal lobes are recruited. This research allows scientists and educators to realise that the brain hemispheres work together rather than separately.[14] However, brain injury or any kind of insult that leads to disorganised brain networks can produce a disorder known as acalculia or dyscalculia (the inability to calculate) in which the brain areas detailed above will not be recruited normally. Specifically, children and adults with this disorder cannot understand the quantity meaning of numbers. For example, they

12. As reported by Dr. Diego Alonso during the Granada forum, both behavioural and brain-imaging studies suggest that the processing of exact arithmetic uses the left frontal lobe (a region usually active during verbal memory tasks). On the other hand, arithmetical estimations involve the left and right inferior parietal lobes (areas associated with visual and spatial tasks). The prefrontal cortex and the anterior cingulate cortex play an important role during complex calculations by controlling nonautomated strategies. It should be noted, however, in mathematical processing that other parts of the brain are involved in addition to the ones noted. On a more speculative note, Dr. Alonso pointed to the work of Dr. George Lakoff and Dr. Rafael Nuñez suggesting further possible work for cognitive neuroscience: to what extent do people use image-schemas (that is to say, spatial relations such as containment, contact, center-periphery, etc.) and conceptual metaphors (cross-domain mappings preserving inferential structures) to create and understand mathematics (see Granada report, OECD website, *op. cit.*)?
13. Dr. Dehaene's brain model describing a system of brain areas active when children are learning or performing arithmetical operations. For further information see Dehaene, S., Spelke, E., Pinel, P., Stanescu, R., and Tsivlin, S. (1999), "Sources of mathematical thinking: Behavioural and brain imaging evidence", *Science*, Vol. 284, No. 5416, pp. 970-974.
14. See: "Hemispheric specialisation", in 4.6.2 below.

would not be able to perform calculations as simple as "3 minus 1" or to understand what number is between 2 and 4. In other words, they have lost the spatial concept of quantity.

Aside from brain damage or disorder, researchers posit that there could be at least two different causes for mathematical difficulties. One possible cause is that some network, like the one that is associated with quantity, may be impaired or disorganised, making it difficult for the person to access information about the inclusion of numbers. Another cause, and probably more common, is that children have yet to learn to connect a quantity representation with both verbal and visual symbols. This can be difficult for children to accomplish, as symbolic thinking or transformations come with experience, both educational and cultural.

As in the previous case of dyslexia, Dr. Dehaene's cognitive neuroscientific model provides a task decomposition of mathematical skill that can be used to devise or validate pedagogical approaches. In particular, the dissociation between the quantity representation and the verbal system supports the possibility of thought without language which means that pedagogical material that emphasises a spatial or concrete objects metaphor for numbers, such as the metaphor of a number line or the Asian abacus, may be particularly well adapted to teach number sense. As an illustration of the efficacy of teaching mathematics by accessing the quantity representation system, the Right Start programme[15] teaches basic arithmetical skills like counting, correspondence between number and quantities, and the concept of the number line. This programme teaches children a spatial analogue of numbers using physical objects like the game of "Snakes and Ladders".[16] This type of training has been successful in remediating children to such an extent that after going through 40 sessions of 20 minutes each, some of these kids started to bypass normally developing children in mathematical class.

4.4. Emotions and learning

4.4.1. *The emotional brain*

In the past, when discussing goals for education, most discussion centred on how to achieve cognitive mastery through reading, writing, and mathematical skills. However, scientists are beginning to realise through experiments what

15. The Right Start Programme – see Dehaene, S. (1997), *The Number Sense*, Oxford University Press, Getty Center for Education and the Arts.
16. This is originally an Asian board game called Parcheesi, and a game of morality (with ladders leading to higher levels of good with snakes impeding access to these higher levels). The game still exists in the West and East, but has since become a game of using one's mathematical skills to win the greatest number of points.

educators have seen in schools: emotions are in part responsible for the overall cognitive mastery present in children and adults and therefore need to be addressed more fully. The education and nurturing of children interact deeply with the development of their brain and of their natural expertise. This is particularly important regarding those aspects of education where emotional competency or flexibility is important. It is also what Dr. David Servan-Schreiber mentions as important in the maturation of a responsible citizen. Some researchers questioned whether the educational system could accommodate the nurturing of emotional competency. Currently, these personality aspects are not addressed in school systems or by educational policies as explicit educational variables on which to concentrate. By discovering the neural basis of these personality variables, cognitive neuroscience can contribute to making them more explicit and understandable potentially, enabling an evolution of educational systems toward the teaching of emotional and self-regulation competency. One important benefit for educational policy would be a greater precision and insight about how this type of self-regulation develops in children and how its development relates to the maturation of underlying neural systems.[17]

Contemporary cognitive neuroscience provides the tools for performing fine-grained componential analyses of the processing that underlies specific tasks. Such analyses have traditionally focused on the cognitive aspects of learning. Similar analyses in the emotional or affective areas have been neglected, as they have not yet been recognised for their role in successful cognitive function. As such, information in this domain is sparse and incomplete. Lack of measurement and theoretical foundation limits progress of the study of emotional regulation in educational practice.

Although the neuropsychological research in emotional regulation is lacking, scientists have established the biological components of emotional expression. Centrally, within the human brain, there is a set of structures collectively known as the limbic system (see Figure 3). The main structures of this system are the amygdala and the hippocampus. This region of the brain has also been referred to as the "emotional brain" and has connections with the frontal cortex. When these connections are impaired either due to stress or fear,[18] social judgement suffers, as well as cognitive performance, because the emotional aspects of learning, including responses to reward and risk, are compromised. As an example of when the interaction between the emotional and the cognitive parts of the brain are impaired: a previously very successful and intelligent (according to traditional criteria: IQ 130) accountant in Iowa, studied by Dr. Antonio Damasio, had of a part of his brain removed due to a lesion; specifically, the communication between the cognitive and the emotional parts of his brain were severed. Following the surgery, he continued to have an IQ well above average for several years, during

17. See the New York report on the OECD website, *op. cit.*

which he was under medical observation. However, his social judgement became so impaired that he lost his job, failed to keep another job, got involved in a number of shadowy business ventures and eventually divorced his wife of 17 years only to remarry a wealthy woman considerably older than him whom he described as "an ageing socialite." This example describes an extreme case of the loss of social judgement. More importantly to the process of education is the fact that this individual still had an IQ well above average *after* the operation.[19]

Figure 3. **Inner structure of the human brain, including the limbic system**

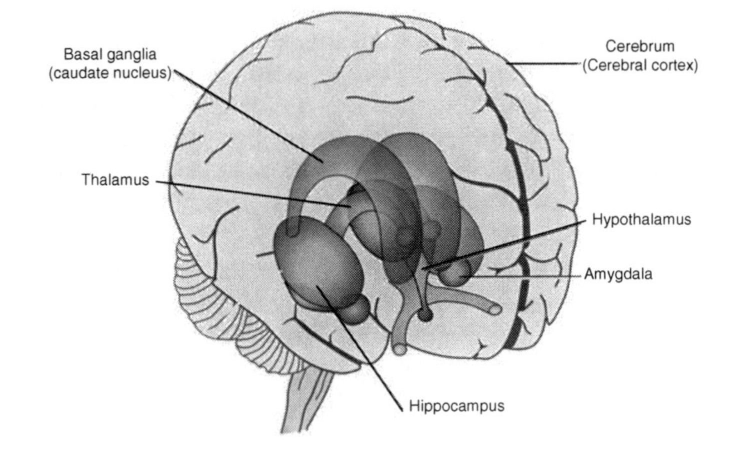

Source: Adapted by Jean-Pierre Souteyrand from Thompson, R.F. (1993), *The Brain: A Neuroscience Primer*, W.H. Freeman and Co., New York.

18. On this specific emotional front, Dr. Joseph LeDoux presented, during the New York forum, recent work on the amygdala. This structure has a critical role in processing the emotion of fear. His research identified specific brain networks of which the emotion of fear is the product. Other emotions may be the products of different brain networks, possibly unrelated to the fear system. The capacity to detect and respond immediately to danger is due to the amygdala (at least in part, as the amygdala also has other emotional contributions). The amygdala interrupts action or thought to trigger a rapid bodily reaction critical for survival. However remote this example may seem from an educational situation, the neuroscience of fear has established several facts that are critical for understanding the role of emotion in education. This function of interruption in the school context may explain some distractibility. Stress and fear in the classroom may impair the capacity to learn by reducing the ability to pay attention to the learning task because of the bodily and emotional demands implicated in the fear system.
19. Recounted by Dr. David Servan-Schreiber. See the New York report on the OECD website, *op. cit.*

All social environments, particularly those of school and work, require that children attain "emotional competency" to function properly. Emotional competency includes, but is not limited to, the ability to be aware of the self, to have self control and compassion, the ability to resolve conflicts and co-operate with others. As pointed out by Dr. Masao Ito, with the emotional brain humans are able to take into account the value of information received, something that distinguishes humans from other mammals.[20]

Psychologically, emotional processing is known to be rapid, automatic, unfiltered by attention and corresponds to what some have described as impulsive. Actually, aspects of this emotional processing make up an individual's temperament or personality. These usually escape normal cognitive instruction but become important when facing a new educational situation. Some children in new situations display fear, some frustration, while others show positive excitation. Research on the amygdala and emotional processing is enabling researchers to understand the complexity of achieving emotional competency as it is related to establishing communication between the emotional and cognitive parts of the brain. With on-going research, neuroscientists are becoming able to demonstrate that emotional processing either helps or hinders the educational process. At least some aspects of such emotional processing may be innate, and not easily changed by environmental circumstances. Thus, one goal of educational procedure would be to discover how to work effectively with students who have different emotional styles.

4.4.2. *Emotional regulation and imagery*

Researchers such as Dr. Stephen Kosslyn are conducting experiments that show that the emotional brain has connections to the perceptive areas of the brain as well. One such part of the brain, the occipital lobe, is not only engaged in perception but also in mental imagery or visualisation. Neuroimaging research has repeatedly shown that the act of imagining or visualising activates many of the same areas of the brain as perception.

Current neuroimaging research in which subjects were visualising aversive stimuli (examples include a battered face or a burned body), showed that certain brain areas were more active than when visualising neutral stimuli (examples include a lamp or a chair). These areas, including the anterior insula (within the

20. According to Dr. Ito (see the New York report on the OECD website, *op. cit.*), one important function from the point of view of education is that "emotional valence" or competency enables the human being, from childhood on, to evaluate a given situation. Our emotional brain makes us more than "mere computers processing information, because it allows us to deal with and to take into account the value of this information, enabling us to have a feeling or a sense for the beautiful."

limbic system) are known to be involved in registering autonomic changes in the body. As Dr. Kosslyn noted, research is just beginning to demonstrate that visualising aversive events not only are registered in the brain, but also affect the body.

What these findings suggest is that learners can alter their emotional state by forming specific mental images. Possible applications of imagery to education include: imagery as a memory aid to remember words better by visualising objects associated with them, and imagery as a mental aid to overcoming test anxiety and phobias. Dr. Kosslyn also mentioned imagery as a hormone regulator. This can indirectly affect cognitive abilities. For example, it is known that the level of the hormone testosterone affects spatial ability. Winning a competition raises the level of this hormone in the blood, while losing a competition lowers it. Thus, it is possible that just visualising such situations can also affect this hormone, which in turn would affect spatial abilities. Work in this area is in progress.[21]

4.4.3. *Effortful control: an educational variable*

Brain research (drawing upon cognitive psychology and child development research) has been able to identify a critical brain region whose activity and development directly relates to the performance and development of self-control. For example, one classic experiment conducted to measure cognitive control is the "Stroop task".[22] In this task, one is shown words that name colours, which are printed in ink that is either the same as the name (*e.g.*, the word "red" in red ink) or different (*e.g.*, the word "red" in blue ink). The subjects are asked to say aloud the colour of the ink, which is much harder if the word names a different colour than if it names the same colour. Performance in Stroop-like tasks tends to activate a very specific region of the brain situated on the frontal midline, just behind the orbito-frontal cortex and called the *anterior cingulate*. The anterior cingulate seems to play a critical role in the brain networks that are responsible for detecting errors and for regulating not only the cognitive processes (as in the Stroop task described above), but also emotions in order to achieve what can be described as the intentional or voluntary control of behaviour.[23]

Self-regulation is one of the most important behavioural and emotional skills that children need in their social environments. This capacity to control one's own

21. See the New York report on the OECD website, *op. cit.*
22. Reported by Dr. Michael Posner during the New York forum (see the New York report on the OECD website, *op. cit.*
23. For a review of the various theories attempting to understand the anterior cingulate's role in the regulation of mental and emotional activity see Bush, G., Luu, P., Posner, M.I. (2000), "Cognitive and emotional influences in anterior cingulated cortex", *Trends in Cognitive Neuroscience*, Vol. 4, No. 6[39], pp. 209-249.

impulses in order to delay gratification is one aspect of the emotional skill "effortful control".[24] From an educational standpoint, paying attention and having voluntary control of behaviour are steps toward success in learning. The role of emotions in education, according to Dr. David Servan-Schreiber, contributes to this "success", which can be defined as a set of loose but meaningful criteria (these criteria include but are not limited to: life satisfaction, establishment of intimate relationships, and the lack of self-provoked life trauma) leading to proactive and rewarding relationships and career prospects. In other words, it contributes to becoming a responsible citizen.

A longitudinal study[25] illustrates the importance of delayed gratification for education. In this study children of 4 years of age were faced with the task of resisting eating one marshmallow displayed before them as they were alone in a (otherwise empty) room in order to get two marshmallows later upon return of the experimenter. The delay of time during which the child succeeded in resisting the impulse to eat the first marshmallow turned out to be significantly correlated with the achievement of later academic success as measured by the ability to deal with frustration and stress, task perseverance, and concentration. In addition, the group of students who exhibited a longer delay of gratification as pre-schoolers turned out to get much higher SAT (Scholastic Aptitude Test – a test incorporating various components of math and reading skills given to adolescents in order to access their potential success and entrance into schools of higher education) scores than those exhibiting a short delay. In fact, there was a significant correlation between delay time and SAT scores.[26]

4.5. The lifelong learning brain

Throughout this section, the concepts of plasticity and cognitive vitality in the older adult serve as the keys to ensure lifelong learning. Research on the adult brain has focused on degenerative diseases and overall cognitive decline. The reasons for this focus range from attempting to help those suffering from diseases and illnesses such as Alzheimer's and senile depression to conducting research where adequate funding is available. Additionally, researchers have found that

24. Dr. Posner mentioned during the New York forum that this concept relates to a child's capacity to self-regulate his/her behaviour both in school and at home. Effortful control can be assessed by synthesising answers from parents to questions about their child's tendency to concentrate on activity (focused attention), to exercise restraint (inhibitory control), to enjoy low intensity stimulation (low intensity pleasure) and to exhibit awareness of subtle change in their environment (perceptual sensitivity) (see the New York report on the OECD website, *op. cit.*).
25. From Dr. Walter Mischel and colleagues.
26. Cited by Dr. David Servan-Schreiber (see the New York report on the OECD website, *op. cit.*).

focusing on degeneration often provides important insights into normal functioning. In this section, decline is discussed along with remediation and rebuilding strategies.

Data[27] show a general decline in most cognitive capacities from age 20 to 80.[28] Declines were noted in tasks such as letter comparison, pattern comparison, letter rotation, computation span, reading span, cued recall, free recall, and so on. By contrast, and in keeping with some of the findings from brain studies earlier in life, there were some notable increases in cognitive capacities across the life-span up to age 70 with some declines by age 80.[29] During the Tokyo forum,[30] Dr. Ito pointed out that there are popular beliefs about decline in the brain with age. Common knowledge asserts that our brain loses 100 000 neurons every day and smoking and/or drinking enhances this loss. However, this belief has been re-examined with new technologies. Dr. Terry and colleagues[31] showed that if one counts the total number of neurons in each area of the cerebral cortex, there is no age dependence. Age-dependence is a factor only when the number of large neurons in the cerebral cortex is counted. These large neurons shrink with the resulting consequence of increasing the number of small neurons, so the aggregate number remains the same. However, there is some decrease of neuronal circuitry as neurons get smaller, and one can expect the number of synapses to be reduced. Moreover, reduced connectivity may mean reduced plasticity, but it does not mean reduced cognitive ability. On the contrary, neural network models have taught researchers that skill acquisition results from pruning some connections while reinforcing others.

Neuroscientists have known for some time that the brain changes significantly over the life-span as a response, in part, to learning experiences. This plasticity[32] or flexibility of the brain to respond to environmental demands is encouraging and is starting to lead researchers to better understand the role of synaptogenesis (the formation of new connections among brain cells) in the adult brain. Moreover, long term learning actually modifies the brain physically because it requires the growth of new connections among neurons. For example, if the reader remembers

27. From the University of Michigan.
28. See also *agingmind.isr.umich.edu/*
29. Tasks included: Shipley Vocabulary, Antonym Vocabulary, and Synonym Vocabulary.
30. See the Tokyo report on the OECD website, *op. cit.*
31. See Terry, R.D., DeTeresa, R., Hansen, L.A (1987), "Neocortical cell counts in normal human adult ageing", *Annuals of Neurology*, Vol. 21, No. 6, pp. 530-539.
32. As Dr. Posner said in his concluding talk at the New York forum, "I think we did bury the myth of birth to three and replaced it with ideas of the importance of both plasticity and periodicity. In other words, the brain is plastic and yet certain things happen at certain periods in our life that are important both for the early years and, of course, for lifelong learning." (see the New York report on the OECD website, *op. cit.*).

anything from this book 6 months from now, it will be because his brain was anatomically modified while reading it (and later, while recalling parts of it).

What cognitive neuroscience has learned is that we must distinguish between synaptogenesis occurring naturally early in life and synaptogenesis associated with exposure to complex environments over the life-span. To illustrate: mastery of grammar appears to occur best at a younger age, but vocabulary learning continues throughout life. Researchers refer to the former as experience-expectant plasticity and the latter as experience-dependent plasticity.[33]

Many researchers believe that experience-expectant plasticity characterises species-wide development; moreover, experience-dependent plasticity is the natural condition of a healthy brain, a feature that allows us to learn right through to old age, and helps account for individual differences in learning.

4.5.1. *Ageing and illness: Alzheimer's disease and senile depression*

Funding for research on the adult and the ageing brain is typically centred on disease models. This fact is explained by the massive and growing cost of neurodegenerative illnesses for society world-wide.[34] In the United States alone, Alzheimer's disease affects approximately 4 million adults and costs the economy about $100 billion per year.

The impact of neurodegenerative disorders is most acutely felt in the area of cognitive function with ageing. Not only may neurodegenerative illnesses rob the individual of his or her sense of self, but they also rob society of accumulated

33. Learning processes that depend on a sensitive period, such as grammar learning, correspond to experience expectant phenomena in the sense that for learning to occur easily, relevant experience is expected to happen in a given time window (the sensitive period). Experience-expectant learning is thought to occur during the early years of life. Learning processes that do not depend on a sensitive period, such as lexicon learning, are said to be experience-dependent phenomena in the sense that the period during which the experience of learning can occur is not constrained by age or time. This type of learning is believed to occur throughout life (see also 4.5.3 below).
34. As noted by both Drs. Raja Parasuraman and Jarl Bengtsson during the Tokyo Forum (see Tokyo report on the OECD website, *op. cit.*). According to Dr. Shinobu Kitayama, ageing cognition must be studied as a function of cultural and social belief systems surrounding the notion of ageing and related ones such as rationality and well-being. The most important project of cultural psychology is to raise questions about the presumed validity of the universality of many mental processes, to show that possible alternatives exist, and to use appropriate analyses to broaden the empirical data base of the human and behavioural sciences. Biological ageing occurs necessarily in a particular cultural context. Depending on the specific nature of the context, it can have quite divergent consequences. The consequences of ageing on cognition must be examined in respect not only to a more holistic, encompassing, relation-centered, and wisdom-based cognition, but merge with a more analytic, object-focused, and individual-centered cognition (see Tokyo report, OECD website, *op. cit.*).

expertise and wisdom. With the ageing of populations, world-wide, this problem will increase.

Dr. Raja Parasuraman explained how Alzheimer's disease is responsible for irreversible brain damage. The symptoms of this disease usually start in late adulthood and involve marked defects in cognitive function, memory, language, and perceptual abilities. The brain pathology associated with Alzheimer's disease is the formation of senile plaques.[35] These changes are particularly evident in the hippocampus, a part of the "emotional brain" crucially involved in short-term memory and in entering new material to be stored in long-term memory.

As there are no reliable methods for detecting Alzheimer's disease, early onset of this disease may be better diagnosed either behaviourally or through genetic testing. It is difficult behaviourally to diagnose early onset, since little is known about the cognitive changes associated with normal ageing. Declining cognitive functions with older age[36] overlap and are similar to those of the preclinical symptoms of Alzheimer's disease. According to some researchers, it may be profitable to direct research resources to the study of attentional functions[37] for early detection of the onset of Alzheimer's disease for at least two reasons. First, attentional functions are found to be impaired even in very mildly affected individuals, providing, perhaps, valuable early warnings. Second, a major area of dysfunction in Alzheimer's disease is in memory function,[38] which can often be addressed by the study of attentional functions.

The neural systems mediating attentional functions[39] are relatively well understood and have been the object of much study. Importantly, two aspects of spatial selective attention (attentional shifting, and spatial scaling) are markedly impaired in the early stages of Alzheimer's disease. Therefore, tasks that assess these functions can serve as useful candidates for early diagnosis. Studies involving event-related brain potential (ERP), Positron Emission Tomography (PET), and functional Magnetic Resonance Imaging (fMRI)[40] indicate that attentional tasks indeed provide sensitive behavioural assays of early attentional dysfunction.

Another approach to early detection of Alzheimer's disease is to begin to identify normally ageing adults who are at *genetic* risk for developing Alzheimer's disease. Recent studies implicate the inheritance of the apolipoprotein E (APOE) gene in the development of Alzheimer's disease.[41] Compared to those without an

35. These are clusters of abnormal cell processes surrounding masses of protein; tangles of neurofilaments inside neurons; deterioration of neuron dendrite's and loss of neurons.
36. See *agingmind.isr.umich.edu/*
37. Dr. Parasuraman during the Tokyo Forum (see Tokyo report on the OECD website, *op. cit.*).
38. Particularly within cholinergic systems.
39. These functions include selective attention, vigilance, and attentional control.
40. For a review of these brain-imaging technologies, see 4.2. above and the glossary.

e4 allele, e4 carriers exhibit spatial attention deficits that are qualitatively similar to those shown by clinically-diagnosed Alzheimer's disease patients: 1) increased attentional disengagement and 2) reduced ability to scale spatial attention. These attentional deficits can appear in otherwise asymptomatic and healthy adults as early as those in their 50s.

Both the behavioural and genetic indicators can lead to the development and testing of new markers for predicting severe cognitive decline in older adults. Equipped with improved diagnostic evidence, pharmacological and behavioural treatment/intervention strategies for enhancing cognitive function in adults can be developed and extended. In a recent series of studies[42] the benefits of attentional cueing (alertness and vigilance training) were shown to reduce the symptomology of Alzheimer's disease by reducing attention deficits and enhancing learning in both healthy adults and Alzheimer's disease patients. Interventions such as these have the possibility of being useful because the fine structure of synaptic connections in the brain is not under direct genetic control but is shaped and reshaped throughout the life-span by experience.

Depression is an illness associated with a host of symptoms including lack of energy concentration, and interest. Symptoms also include insomnia, loss of appetite, and anhedonia (incapacity to experience pleasure). Depression in older adults, unlike that found in younger people, presents a more complicated etiology and is thus more difficult to treat.[43] As with other age-related disorders, senile depression imposes major health and societal burdens. Currently, depression in the elderly is the second most frequent mental disease following dementia.[44]

Depression can be caused by a general degeneration of the brain (*e.g.*, Alzheimer's disease, Parkinson's disease, and stroke). A major difference between depression in the elderly and in younger people is that there appears to be less of a genetic contribution to the disease in the elderly. In addition to the organic causes noted above, depression in the elderly can be traced in some cases to the

41. The APOE gene is inherited as one of three alleles, e2, e3, and e4, with the e4 allele associated with greater risk of developing Alzheimer's disease. See Greenwood, P.M., Sunderland, T., Friz, J., and Parasuraman, R. (2000), "Genetics and visual attention: Selective deficits in healthy adult carriers of the e4 allele of the apolipoprotein E gene", Proceedings of the National Academy of Sciences, United States, Vol. 97, pp. 11661-11666.
42. See presentations by Drs. Parasuraman and Kramer, Tokyo Forum (see Tokyo report on the OECD website, *op. cit.*).
43. Dr. Shigenobu Kanba mentioned during the Tokyo forum that older adults tend to have complicated cases of depression due to physical declines such as cardiovascular-type blockages in deep brain areas, also known as micro infarctions in basal ganglia (see Tokyo report on the OECD website, *op. cit.*).
44. Dr. Kanba, Tokyo Forum (see Tokyo report on the OECD website, *op. cit.*).

sudden loss of social roles, the loss of important and close people, and a decline in economical, physical and psychological capacity.

According to Dr. Shigenobu Kanba, it is important for society to recognise and respond to two important aspects of senile depression that can be ameliorated. The first is primarily psychological and associated with "object loss". For example, society should work to see that old people are not suddenly deprived of their social roles, employment, or self worth. One way of doing this is to provide ways in which contributions of the aged to society[45] can be celebrated and used.[46]

Not only are practical problem solving and personality openness contributions the elderly can make to society, but they are also now known to be positively correlated with creativity and well being. Indeed, there long has been evidence that creativity is largely separate from intelligence (one needs only a certain "threshold" amount of intelligence to be creative, but above that amount there is no relation between the two capacities). Therefore, any age-related decline in higher-order cognitive functioning does not necessarily affect creativity. Dr. Yoshiko Shimonaka conducted a study in order to examine the effects of ageing on creativity in Japanese adults ranging in age from 25 to 83.[47] No age differences were found on fluency, originality of thinking ability, productivity and application of creative ability.[48] However, gender differences were found on fluency and productivity with women outscoring men. These results suggest that

45. Contributions include, but are not limited to, the use of knowledge, expertise and maturity.
46. In this regard, the spontaneous learning communities in Australia described during the Tokyo forum by Dr. Denis Ralph provide an example of how an attentive community can assist the elderly. In these communities, the elderly, with the support of modern computers rediscover learning. (*Australian National Training Authority, National Marketing Strategy for Skills and Lifelong Learning*; report presented to Ministerial Committee, November 1999, URL: *www.anta.gov.au*. This report discusses attitudes and values to learning within the Australian community, motivation for involvement in learning and factors likely to influence participation in learning.)
47. Creativity tests developed by J.P. Guilford. Tests consisted of thinking ability (fluency, flexibility, originality and elaboration) and creative ability (productivity, imagination and application).
48. As reported by Dr. Tudela during the Tokyo forum (see Tokyo report, OECD website, *op. cit.*), in the literature on skill learning in normal ageing there are many tasks in which young adults and older adults are directly compared. Within this task set, in some cases young adults perform better than older adults; other times their performances are equal. Unfortunately, progress in the understanding of skill acquisition in normal ageing is hampered because a component processes approach to the analysis of the nature of perceptual, motor and cognitive skills is lacking. A good analysis in terms of component processes and if possible in terms of neural mechanisms is needed. In other words, it is necessary to actively apply the analytic approach of cognitive neuroscience to the design of tasks used in this form of research.

various abilities of creativity are maintained throughout the adult years. Encouraging the elderly to provide guidance to younger people, a process that could prove mutually beneficial, may therefore ameliorate psychologically based depression in the elderly.

4.5.2. *Fitness and cognitive vitality*

The idea that relates physical and mental fitness is an ancient one, expressed in Latin by the poet Juvenal as *"mens sana in corpore sano"*.[49] A review of the animal literature,[50] found reason for optimism in the enhancement of cognitive function.[51]

A recent Japanese study,[52] reported by Dr. Itaru Tatsumi, compared language proficiency of young and elderly Japanese adults (elderly people often complain about a difficulty in the retrieval of proper names of their acquaintances and of the names of famous people). Young and elderly subjects were asked to say aloud as many words in a given category as they could (for semantic and phonological categories) within 30 seconds. The number of words the elderly subjects could retrieve was approximately 75% of the number retrieved by the young subjects, showing less word-fluency for the elderly. Additionally, retrieval of famous names was difficult (their mean performance was about 55% to that of young subjects). Going beyond psychophysical studies, Dr. Tatsumi reported on a PET activation study of elderly and young subjects observed during word-fluency tasks. In young subjects, during word retrieval of proper names, the left anterior temporal lobe and frontal lobe were activated. During the retrieval of animate and inanimate names, and syllable fluency, the left infero-posterior temporal lobe and left inferior frontal lobe (Broca's area), were activated. By contrast, activated areas in the elderly subjects were found to be generally smaller than those in the young subjects, or sometimes inactive. Moreover, areas that were not active in the young subjects were activated in the elderly. Conclusions based on these findings are subject to further investigation as one could interpret the latter activations to reflect an effort to compensate for deficient word retrieval in the elderly subjects. Another conclusion, in favour of the vitality of the ageing brain, is that fluency or experience with a task necessarily reduces activity levels. Presumably due to higher processing efficiency and use of brain mechanisms, these tasks can also be shuttled to different areas of the brain for processing.

49. "A sound mind in a sound body."
50. Conducted by Dr. Kramer.
51. These studies looking at (among other things) synaptogenesis and neurogenesis, and positive biochemical changes associated with brain-derived neurotrophic factor (BDNF), dopamine receptor density, and choline uptake.
52. By Sakuma *et al.* (see Tokyo report, OECD website, *op. cit.*).

A recent reanalysis[53] of existing longitudinal data using meta-analytic techniques suggests a more positive and robust association between enhanced fitness and cognitive vitality, particularly in executive processes (management or control of mental processes). Emerging data suggest that the regions of the brain associated with executive processes (*e.g.*, the frontal cortex and hippocampus[54]) show large and disproportionate age-related declines. These declines may be slowed by physical fitness. In particular, task improvement was shown to be positively correlated with cardio-vascular function.[55] Specific training studies also show that there are positive results for spatial orientation, inductive reasoning, and complex task-switching activities such as driving. In general, there is tentative, but growing, evidence that behavioural, non-pharmacological, interventions including enhanced fitness and learning (following the slogan "jog your brain!") can contribute to improvements in performance even into old age. The applicability of these results outside of the laboratory will be the focus of further research.

4.5.3. *Plasticity and lifelong learning*

The brain's ability to remain flexible, alert, responsive and solution oriented is due to its lifelong capacity for plasticity. At one point, neuroscientists thought that only infant brains were plastic. This was due to the extraordinary growth of new synapses (synaptogenesis) paralleled with new skill acquisition. However, non-human primate and human primate data, uncovered over the last two decades, has confirmed that the brain retains its plasticity over the life-span.[56] Moreover, parts of the brain, including the all important hippocampus, have recently been found to generate new neurons throughout the life-span.

53. This was done to lessen the methodological limitations inherent in the three other types of methods used in human research: cross-sectional, epidemiological, and longitudinal studies. The cross-sectional studies suggest large and robust cognitive benefits for fit older adults, but these studies suffer from the typical limitations of cross-sectional studies (self-selection bias). Epidemiological studies list a number of factors associated with cognitive vitality including strenuous exercise, making it difficult to separate out the contribution of individual factors. Findings from longitudinal studies are mixed with six studies finding cognitive benefits associated with enhanced fitness, four showing no relationship, and two with ambiguous results.
54. Raz, N., Williamson, A., Gunning-Dixon, F., Head, D. and Acher, J.D. (2000), "Neuroanatomical and cognitive correlates of adult age differences in acquisition of a perceptual-motor skill", *Microsc Res Tech*, Oct. 1, Vol. 51, No. 1, pp. 85-93.
55. Dr. Kramer stressed during the Tokyo forum that adults used in this study are normally exercising adults, who have had regular exposure to exercise over time (see Tokyo report on the OECD website, *op. cit.*).

Research by Dr. McCandliss and others has shown that older brains can adapt in order to overcome barriers in language processing and reading, used here as an example. These new findings are leading to:

- A better understanding of the different ways the brain can process language.
- Clearer ideas on how children and adults can naturally overcome language-processing obstacles (especially in dyslexia).
- Insights into how strategies may recruit different neural networks and help those with reading and speech disabilities.
- Ways to help second-language learners (both adults and children) improve their phonological understanding of the language.

Much of the research in ageing is driven by models of disease and pathology. As such, it is important to separate the responsibilities of the health-care industry from those of education in this context, particularly with regard to syndromes such as Alzheimer's disease and senile depression.[57] Gains for both sectors of society will flow from

56. Progress on the effects of the environment on learning and concomitant insights into brain plasticity may be helped by narrowing the focus to particular brain regions and specific learning skills. For example, the hippocampus is known to be involved in processes of spatial memory and navigation [see Burgess, N. and O'Keefe, J. (1996), "Neural computation underlying the firing of place cells and their role in navigation", *Hippocampus*, Vol. 6, No. 6, pp. 749-762]. Intriguing research comparing London taxi drivers with non taxi drivers suggests a strong relationship between the relative size and activation of the hippocampus and successful navigation, a relationship that appears to have a distinct temporal quality [see *i)* Maguire, E.A., Frackowiak, R.S. and Frith, C.D. (1996), "Learning to find your way around: A role for the human hippocampal formation", Proceedings for the Royal Society of London (B), *Biological Sciences*, Vol. 263, pp. 1745-1750; *ii)* Maguire, E.A., Frackowiak, R.S. and Frith, C.D. (1997), "Recalling routes around London: Activation of the right hippocampus in taxi drivers", *Journal of Neuroscience*, Vol. 17, No. 18, pp. 7103-7110; *iii)* Maguire, E.A., Gadian, D.S., Johnsrude, I.S., Good, C.D., Ashburner, J., Frackowiak, R.S. and Frith, C.D. (2000), "Navigation related structural changes in the hippocampi of taxi drivers", Proceedings of the National Academy of Sciences, United States, Vol. 97, No. 8, pp. 4398-4403]. Similarly, there is a positive correlation between the enlargement of the auditory cortex and the development of music skill [see Pantev, C., Osstendveld, R., Engelien, A., Ross, L.E., Roberts, L.E., and Hoke, M. (1998), "Increased auditory cortical representation in musicians", *Nature*, Vol. 392, pp. 811-814], and also for motor areas and finger movements even for durations as short as five days for adults [see Pascual-Leone, A., Nguyet, D., Cohen, L.G., Brasil-Neto, J.P., Cammarota, A. and Hallett, M. (1995), "Modulation of muscle responses evoked by transcranial magnetic stimulation during the acquisition of new fine motor skills", *Journal of Neurophysiology*, Vol. 74, No. 3, pp. 1037-1045]. On the other hand, patients with Parkinson's disease (which involves abnormal functioning basal ganglia) are unable to learn some new skills [see Gabrieli, J.D., Brewer, J.B., and Poldrack, R.A. (1998), "Images of medial temporal lobe functions in human learning and memory", *Neurobiology of Learning and Memory*, Vol. 20, No. 1-2, pp. 275-283].

57. Dr. Ito during the Tokyo forum (see Tokyo report on the OECD website, *op. cit.*).

pathology-centred research agendas: current research agendas in ageing are not driven by concerns about learning. Agendas in ageing that are focused on learning can draw some measure of reassurance concerning the plasticity of many brain functions and the robustness of some cognitive functions well into old age. With age, some cognitive abilities do decline, especially with the onset of disease. However, with improved imaging technologies and more sensitively planned research protocols, research into remediation strategies (including learning tasks) and extended cognitive function will serve to broaden both satisfaction and potential contribution in the older adult.

4.6. Neuromythologies

4.6.1. *Separating science from speculation*

With the advent of functional imaging technology, cognitive neuroscience is beginning to produce important research on the neural foundations of cognitive performance. Current research results have sparked a tremendous amount of commentary and speculation among scientists, researchers, education specialists, and policy-makers. Since such research proves to have merit, many want to know how educational practice can be improved or enriched by the application of these research findings.[58] As a result of both pressure to improve overall school performance and excitement and interest about education that could be brain-based, many myths and misconceptions have arisen around the mind and brain outside of the scientific community. Teachers and educational specialists are eager to put into practice what they have read in the popular press,[59] and policy-

58. Not only educational practices, but also everyday parenthood may benefit from research findings. In fact, parents are an important "market" for neuromythologies.

59. "(...) this brain information is on the television, in the newspapers, in the magazines: what does it mean for the classroom teacher?" (Mark Fletcher, during the Granada forum). "[The teachers] hear a lot about their subjects, about mathematics or biology, or whatsoever, but they really have a big lack in neuroscientific and psychological learning theories. I think we should look into this direction and ask what teachers could learn from cognitive neuroscience." (Dr. Heinz Schirp during the Granada forum.)
Teachers, of course, are not neuroscientists, but it is both understandable and desirable that they look to the work of neuroscientists to help them improve teaching. Given that those who promulgate brain-based education to teachers fail to also convey the relative paucity of research to support their claims, teachers might be tempted to too readily adopt so-called "brain-based" teaching strategies that are in fact not based on any evidence at all. The scientific community should be sensitive to these issues. A challenge, therefore, is to strengthen pedagogical knowledge and strategies by inviting teachers to *a)* share their knowledge among themselves and *b)* share this knowledge with the neuroscientific community. Thus, the neuroscience community will be able to ground some of its research questions within the authentic experiences of good teachers. Hence, it is necessary to educate the public about both the gains due to cognitive neuroscience, but also the need to focus on "simple" questions about elementary processes first. There is of course a lot of work to do in order to integrate insights about elementary processes into the complex context faced by educators. Furthermore, educators can play a key role in helping identify such questions, that might be tractable for neuroscientists (see the Granada report on the OECD website, *op. cit.*).

makers want to enact effective educational policy by using research-based information. Even business is eager to commercialise on what is perceived of financial interest in brain-based learning tools. Due to the expectations of the applicability of brain research to educational practice, myths have rapidly developed and range from the benefit of synaptogenesis, to hemisphere dominance, to critical periods of learning and enrichment – to name the most popular ones. When misconceptions such as these are both argued for and criticised in journals and the popular press, educators and policy-makers alike are left in a quandary discerning fact from fiction. Although some myths do have some truth to them, careful reading of the original research from where they came from demonstrates that this research has often either been misinterpreted (simplified) or based exclusively on animal studies with limited implications for human beings.[60]

In the past, most scientists have claimed that at birth, the human brain has all the neurons it will ever have. However, with the advent of new technologies, this fact is being challenged. Some mechanisms, such as those that control our basic survival instincts are in place at birth, but most of the new-borns mental circuitry results from experiences – how and when these connections are formed is a subject of debate. Some scientists argued that these circuits are completed by age 3, others believed that they continue until adolescence; more recently, a consensus seems to emerge, implying that synaptic connections are formed throughout the life cycle. This emerging and recent consensus can have profound implications as to the way the education system is organised.

The goal of this section is to explain the origin of some of the more prevalent myths that the public has, to highlight why they are detrimental and/or non-effective for educational practice and to discuss how best to interpret scientific data.

The decade 1990-2000 was declared the "Decade of the Brain" in the United States. At the same time, world-wide research on the cognitive and emotional functioning of the brain has been stimulated.[61] Although much of this research has been of very high quality, some of its findings have been over-interpreted in terms of their implications for learning. Such examples are presented below.

60. While animal studies have proved essential and necessary in understanding some aspects of human development, caution must be exercised when applying results of experimental data to human learning and cognition. More generally, history shows that establishing parallels between animal and human behaviour without exercising extreme caution can prove misguided, if not dangerous.
61. This is just one example where political vision has encouraged scientific research and, beyond, started to influence educational change through appropriate funding.

Neuroscientific results must be taken as preliminary in nature for several reasons:

- Their statistical results might not be of the highest relevance (subtraction method and averaging[62]).
- Results on the same subject can differ due to methodological and theoretical considerations.
- The laboratory setting might not be the appropriate place to test a skill as it is an unnatural and contrived setting.
- A single study cannot justify a certain classroom strategy.
- In the popular press, in order to appeal to the greatest number of people, often the reporting of brain research is over-simplified; this is the origin of almost all misconceptions and misunderstandings about science.

Some current claims about the neuroscientific basis for learning must be approached with a healthy dose of skepticisim. Current and emerging technologies produce both interesting and promising results, but these will prove even more relevant and useful for education if previous misconceptions and misbeliefs about science are eradicated.

The genesis of a neuromyth usually starts with a misunderstanding, a misreading and in some cases a deliberate warping of the scientifically established facts to make a relevant case for education or for other purposes. There are three popular myths discussed in this chapter: hemisphere dominance or specialisation, synaptic development and learning, "critical" periods and enrichment (including the myth of birth to three).

62. These methods are considered somewhat weak, because comparing two different results that might share some elements in common will not clarify the differences between the two results. In neuroimaging data, for example, condition A is one task and condition B is another, different, task. In order to find the differences between them to ascertain which condition activates a particular brain area, the subtraction method is often used. This consists of looking at all the activation points on one image due to condition A and then looking at all the activation points on another image due to condition B. If the two conditions are really different, one can reason that subtracting one image from another will show only those areas in the brain pertinent to a particular condition. The problem with this method is that from one condition to another, the brain does not necessarily stop its activation from the previous condition just because the condition has ended (sometimes there's a residue of activation) and sometimes both conditions will activate the same brain areas. So defining with certainty, which brain areas are activated by a particular condition is not always accurate. Using the same example of condition A and B, the averaging method involves taking data from different subjects, for example, from the same condition and averaging together the results. The problem with this method is that even if the individual results are greatly varying (which they often are), the effects of what could be significantly different are lessened, thereby reducing potentially problematic results and generating inaccurate conclusions.

4.6.2. *Hemisphere dominance or specialisation*

An example of a misconception about brain science and learning concerns "right brain *versus* left brain learning". The claims about brain hemisphere specialisation and its relationship to learning point out a central shortcoming in the brain-based learning movement. It is generally asserted by non-specialists that the left hemisphere is the logical one and codes for verbal information, while the right hemisphere is the creative one and codes for visual information. Often, these ideas become polarised over time, and attributes of the brain are thought to come from either one hemisphere or the other. These attributes are then substituted for character traits making people claim, for example, that artists are "right brained" while mathematicians are "left brained".

Although Dr. Dehaene has completed an analysis[63] which demonstrated the left hemisphere's responsibility in the processing of number words (*e.g.*, "one", "two"), he also showed that both right and left hemispheres were active in the identification of Arabic numerals (*e.g.*, "1", "2"). Similarly other recent data show that when the processes of reading are analysed into smaller components, subsystems in both brain hemispheres are activated (*e.g.*, decoding written words or recognising speech sounds for higher-level processes such as reading text). Indeed, even a quintessential "right-hemisphere ability", encoding spatial relations, turns out to be accomplished by both hemispheres – but in different ways. The left hemisphere is better at encoding "categorical" spatial relations (such as above/below, or left/right) whereas the right hemisphere is better at encoding metric spatial relations (*i.e.*, continuous distances). Moreover, neuroimaging has shown that even in both of these cases areas of the two hemispheres are activated, and these areas work together. The brain is a highly integrated system; one part rarely works in isolation.

There are certain tasks, such as face recognition and speech production, which are dominant to one particular hemisphere, but most tasks require both hemispheres to work in parallel. This is an example of how certain and rather limited research findings turn into well-known neuromyths.

Asking a few questions prior to accepting published results as appropriate for education practice is necessary. Some general questions to reflect upon include:

- Is this an isolated case or are there others to support the claims being made?
- Are the studies describing events or are they testing hypotheses?
- Is the learning task used appropriate for the population tested? In other words, would this be an appropriate task for teaching school-aged children?

63. Using word masking and unconscious priming.

4.6.3. *Synaptic development, "enriched" environments and "critical" periods*

Neurons, or brain cells, are the foundation of the human brain. These cells communicate with one another via synapses, or junctions, where nerve impulses travel from cell to cell and support skill development, learning capacity, and growth in intelligence. At birth, the number of synapses is low compared with adult levels. After two months and peaking at ten months, the synaptic density in the brain tissue increases exponentially and exceeds adult levels. There is then a steady decline to (and stabilisation at) adult levels around age 10.

The process by which synapses are being created in great numbers during normal periods of growth is called synaptogenesis. It varies across the life-span with differential growth periods for different brain areas, contingent upon experience. The process by which synapses decline is referred to as "pruning" and is known to be a normal and necessary process of growth and development. In general, over the life-span, synaptic densities follow a skewed Gaussian curve with a sharp increase seen in infancy, a levelling off during adulthood and a slow decline in very old age.

In laboratory experiments with rodents, presented in New York by Dr. William Greenough, synaptic density has been shown to increase by the addition of a complex environment. A complex environment was defined in this case as a cage with other rodents, and various objects to explore. When these rats were subsequently tested on a maze learning test, it was demonstrated that those rodents as compared with a control group living in "poor or isolated" environments, performed better and faster in the maze learning task.[64] The conclusion was made that rats in "enriched" environments had increased synaptic density and thus were better able to perform the learning task.

This is the beginning of a neuromyth. Even if synaptogenesis and synaptic pruning are likely to have important learning implications for rodents, it is not proved that the same holds true for human beings. This rigorous, scientifically established experimental data on rodents has been combined by non-specialists with basic human development to assert that educational intervention, to be most effective, should be timed with synaptogenesis. The neuromyth logic is that the more synapses available, the higher the potential nerve activity and communication, thus making better learning possible. An associated belief is that early educational intervention using "enriched environments" can save synapses from pruning, or can create new synapses, thereby leading to greater intelligence or greater learning capacity. Feeding this is the additional problem of quoting the facts of a pertinent study and then assigning meaning that goes well beyond the evidence presented in the original research paper.

64. Diamond, M. *et al.* (1987), "Rat cortical morphology following crowded-enriched living conditions", *Experimental Neurology*, Vol. 96, No. 2, pp. 241-247.

Apart from the descriptive data on synaptic activity and therefore density, described above, there is not yet much neuroscientific evidence in humans about the predictive relationship between synaptic densities early in life and improved learning capacity. As Dr. John Bruer has repeatedly asserted,[65] studies on this cannot yet form the foundation for principles about how to improve formal education. However, this does not mean that brain plasticity in general, and synaptogenesis in particular, are irrelevant for learning, but further research is needed.

As could be predicted, any claim based on improper deductions and generalisations from an often misunderstood conception about synaptogenesis/synaptic pruning has its weaknesses. First, it is still difficult to obtain direct concurrent evidence relating counts of synaptic densities to learning. Up until recently, these data have been collected from humans or animals posthumously. Second, there is not yet much neuroscientific evidence in humans about the predictive relationship between synaptic densities early in life and densities later in life. Third, there is no direct neuroscientific evidence in either animals or humans linking adult synaptic densities to greater capacity to learn.[66] The point of this critique is not to condemn early educational interventions, but rather to challenge the claim that the value of early educational intervention is based on a neuroscientific consensus or brain imperative.

Considering the popular myth of "synaptic development and learning", it is wise to ask some questions: Is the study backed up by scientifically valid research? Has the study been replicated in order to arrive at consensus? Did the study or studies actually involve learning outcomes or are the implications for learning claims speculative? In general, did the study or studies rigorously test clear hypotheses or were they largely descriptive in character? How plausible is the chain of causal reasoning from the neuroscientific data to implications for learning? Of what population is the sample representative and to what population do the claims apply?

If in rodents it has been concluded that a complex environment causes increased synaptic density, and rats with more synapses[67] are thought to be smarter than their counterparts who have lived in impoverished environments

65. Bruer, J.T. (1998), "Brain science, brain fiction", *Educational Leadership*, Vol. 56, No. 3, pp. 14-18; Bruer, J.T. (1999), "Education and the brain: A bridge too far", *Educational Researcher*, Vol. 26, No. 8, pp. 4-16; Bruer, J.T. (1999), "In search of brain-based education", *Phi Delta Kappan*, Vol. 80, No. 9, pp. 648-657.
66. Bruer, J.T. (1999), "In search of brain-based education", *op. cit.*
67. Rats raised in the complex (more natural) environment had 20 to 25% more synapses per neuron (what was measured was the ratio of the density of synapses to density of neurons) in their upper visual cortex than rats raised in the deprived environment. The increase in the number of synapses per neuron was accompanied by a change in the number of blood vessels (responsible for transferring nutrients from the blood to the neurons) and in the number of other cells called astrocytes (which have a role in the metabolic support of neurons and in the growth of new synapses between them). In other words, both neural and non-neural tissue was embellished by experience.

(with presumably fewer synapses), then by analogy, the belief has arisen that providing stimulating environments for students will increase their brain connectivity and thus produce better students. Recommendations have been suggested that teachers (and parents) should provide a colourful, interesting and sensory meaningful environment to ensure a bright child.[68]

For over thirty years, neuroscientists have been collecting data about sensitive periods in biological development. As noted earlier, a sensitive period[69] is defined as a time frame in which a particular biological event is likely to occur

68. Arguing from the data on rats about the need for "enriched environments" for children is unjustified (*e.g.*, listening to Mozart, looking at coloured mobiles), particularly considering that parallel neuroscientific studies of the affect of complex or isolated environments on the development of human brains have not been conducted. On the other hand, the rat studies suggest that there is a critical threshold of environmental stimulation below which brain development may suffer. Recent studies of Romanian orphans demonstrate the ill effects of severely restricted environments, but even in these cases, rehabilitation is possible [see O'Connor, T.G., Bredenkamp, D. and Rutter, M. (1999), "Attachment disturbances and disorders in children exposed to early severe deprivation", *Infant Mental Health Journal*, Vol. 20, No. 10, pp. 10-29].
    Other problems with carelessly using this research for educational purposes lie in the following:
    - In the wild, rats naturally live in stimulating environments (drainage pipes, waterfronts, etc.) and so presumably have exactly the number of synapses needed to survive. It doesn't make sense to put them in an impoverished environment because this is an artificial setting, which is unrealistic. So if you put rats in an artificially impoverished environment, their brains will have exactly the density of synapses appropriate for that environment. In other words, they will be just as "smart" as they need to be for living in a laboratory cage. If the same line of reasoning applies to human beings (which is likely, but still has to be demonstrated), given that most humans are raised in normally stimulating environments, their brains are uniquely adjusted for their particular environments.
    - There are too many factors to take into account when defining what an "enriched" environment should look like for the majority of students.
    - The density of synapses has not been shown experimentally to affect mastery of educational skills.
    - Most children naturally grow up in environments that are stimulating. Research has shown that even children growing up in what could be traditionally defined as an impoverished environment (such as a ghetto), may continue, over time, to excel in school and go on to receive degrees in higher education.
69. Sometimes referred to as "critical period"; both terms are often used interchangeably. However, there are subtle differences. "Critical period" implies that if the time frame for a biological milestone is missed, the opportunity is lost. "Sensitive period", on the other hand, implies that the time frame for a particular biological marker is important, but not necessary in the achievement of a particular skill. Mastery can occur, but with more difficulty. Since "critical periods" seem to belong to the popular misconceptions about neuroscience, throughout this document, "sensitive period" will be used to refer to this phenomenon, except when explicitly referred to the misconception.

best. Most of the research was centred on the visual system, primarily in cats and later in monkeys.[70] Past research has shown that blindness in kittens will occur if denied visual stimulation within the first 3 months of life. Misusing scientific data on synaptogenesis, another popular misconception states that from birth to 3 years of age, children are the most receptive to learning. As a consequence of this, the belief among many non-specialists is that if a child has not been exposed "fully and completely" to various stimuli, it will not "recuperate", later on in life, these capacities "lost" in early age. Being exposed to rich and diverse stimuli is what is typically considered an "enriched" environment. However, referring back to the original literature, it should be noted that the data on sensitive periods for cat vision are not simple or always consistent. There are data to suggest that some recovery in vision is possible depending on the length of the deprivation and the circumstances following the deprivation. In other words, it is the balance and relative timing of stimulation that matter, and not that increased or "enriched" stimulation during a sensitive period make for better vision.[71] This misconception uses the previous popular beliefs about synaptogenesis and so-called "critical periods" to make a claim that for full learning to occur, rich diversity and early exposure are best; in fact, early exposure may be just fine, but the claims do not (yet?) have a basis in cognitive neuroscience.

There is a distinction to be made between synaptogenesis occurring naturally early in life and synaptogenesis associated with exposure to complex environments over the life-span. For example, data does seem to suggest that grammar learning occurs best (*i.e.* faster and easier) at a younger age (before age 16, more or less), but that vocabulary learning improves throughout life. Learning processes that depend on a sensitive period, such as grammar learning, correspond to "experience-expectant" phenomena in the sense that for learning to occur easily, relevant experience is expected to happen in a given time frame (a sensitive period). Experience-expectant learning is thought to occur best during certain periods of life. Learning processes that do not depend on a sensitive period, such as lexicon learning, are said to be "experience-dependent" phenomena, in the sense that the period during which the experience of learning can occur is not constrained by age or time. This type of learning can, and does, improve over the life-span.

Sensitive periods do exist, and could over time be useful for education and learning practice, as pointed out by Dr. Hideaki Koizumi, who suggests that "a reorganisation of the education system according to the sensitive periods of the

70. Hubel, D.H., Wiesel T.N. and LeVay, S. (1977), "Plasticity of ocular dominance columns in monkey striate cortex", *Philosophical Transactions of the Royal Society of London* (B), Vol. 278, pp. 307-409.
71. Bruer, J.T. (1998), *op. cit.*, p. 16.

brain", once these are clearly identified, would be desirable. "The most important goal in education seems to be to develop a learning capability suitable for each individual according to the sensitive periods of acquiring cognitive functions. Some basic education should be employed while the brain possesses a high plasticity; in other words, the early stage of education is important. This was known a long time ago in terms of music and language education. The progress of cognitive neuroscience, however, is leading us to further findings. The human brain functions, based upon various functional areas, consist of many modules and frames.[72] Each function module or frame would have a different sensitive period due to the plasticity of neuronal networks. (...) Although education at an early age is highly important, it does not mean that a large part of a person's education must be concentrated into the childhood years. An optimal arrangement of educational items based upon the sensitive periods is likely to be much more effective. Educational items whose sensitive periods occur later in life should be dealt with later." Thus, the neuromyth that the most sensitive period for learning is in the early years of life needs to be revised in the light of recent neuroscientific research, showing that certain forms of learning improve over the life cycle. To summarise, Dr. Koizumi suggests to "reorganise the educational system within the near future by applying recent findings in developmental cognitive neuroscience".[73]

72. See Foder, J.A. (1983), *The Modularity of Mind*, MIT Press, Cambridge; and also Koizumi, H. (1997), "Mind-morphology: an approach with non-invasive higher-order brain function analysis", *Chemistry and Chemical Industry*, Vol. 50, No. 11, pp. 1649-1652.
73. Koizumi, H. (1999), "A Practical Approach to Trans-Disciplinary Studies for the 21st Century – The Centennial of the Discovery of Radium by the Curies", J. *Seizon and Life Sci.*, Vol. 9, No. B 1999.1, pp. 19-20.

Part III

CONCLUSION

Chapter 5

The Way Ahead

5.1. Towards a new learning science based on a trans-disciplinary approach?

To make an end is to make a beginning.
The end is where we start from.

T.S. *Eliot*

"In the past, the trans-disciplinarity, bridging, and fusing concepts that brought together widely divergent fields were the exclusive privilege of genius, but in the 21st century these tools must become more widely available. (...) The provision of a trans-disciplinarian education that will enable future trans-disciplinary studies is an urgent need that we must satisfy for the benefit of future generations."[1]

H. *Koizumi*

Brain-based learning is not a panacea that will solve all education's problems. However, research directed towards understanding learning from this perspective does offer some direction for education specialists, policy-makers and practitioners, who want more informed teaching and learning. It offers better options for students and adults alike who have difficulty with learning.

Following the success of the OECD-CERI project entitled "Learning Sciences and Brain Research" (1999-2001), it has been decided to launch a second phase of this programme (since this decision was made, the three fora held in 2000-2001 are referred to as "Phase 1" of an ongoing project). Phase 2, scheduled over a three-year period (2002-2005), will focus on three research areas which have particular relevance for policy (in terms of curriculum design, teaching practice, identification of individual learning styles) and high potential application. Its goal is to

1. Koizumi, H. (1999), "A Practical Approach to Trans-Disciplinary Studies for the 21st Century – The Centennial of the Discovery of Radium by the Curies", J. *Seizon and Life Sci.*, Vol. 9, No. B 1999.1, p. 7.

both corroborate past research and, more importantly, to further extend current research results. It is anticipated that research coming from selected institutions and high-level scientists will not only foster further hypothesis generation and testing, but will additionally provide new ways in which children, adolescents, and adults can be educated. The three research areas identified for Phase 2 are "brain development and literacy", "brain development and numeracy" and "brain development and learning over the life cycle".

This activity is a work in progress. It would be premature at this stage to draw firm conclusions about precisely how the brain works, how people learn best, and what educational provision can best help them. But it would also be wrong to suggest that questions of this magnitude and significance may not be substantially clarified and partly answered in the near future. There is no doubt that "learning and the brain" must be high on the agenda in OECD countries now and in the years to come.[2] It implies:

1. the promotion of trans-disciplinary relationships;
2. the investment in trans-disciplinary research; and
3. the recognition of the emergence of a new science of learning and the need to develop "institutions of learning science" to facilitate both 1) and 2).

The idea of relatedness provides both an image for the growing understanding of the way in which different parts of the brain co-operate, and a challenge to those who seek to promote and develop this understanding. Science seems to make progress by alternately pursuing two contradictory strategies – distinction

2. A risk here is that a sort of "neuro-divide" develops, not unlike the "digital divide" (currently serving to refer to the limited, if not more, access of the underprivileged to information and communication technologies), in which the privileged would gain first and best access to the fruits of a science of learning. A major challenge, therefore, is to publicise on an ongoing basis the best educative research involving cognitive neuroscience, not simply in research journals, but to help the general public to understand the policy and learning implications of the new work. New forms of information dissemination (*e.g.*, television documentaries, multimedia www sites) should be recruited in this effort. Moreover, scientific findings, in almost all disciplines, raise more and more ethical questions, to be dealt with, at least in democracies, in the political arena. As Dr. Alain Michel pointed out in Granada: "(...) we all know that scientific results can be badly used; we all think about the danger of eugenics, for example. If we are to use neuroscientific research results in the field of education, we will need a code of ethical conduct, so that some excesses can be avoided. Remember, as Rabelais said, in the 16th century: 'Science sans concience n'est que ruine de l'âme'". It is necessary to encourage a large international debate about all the ethical questions addressed in this framework. As Dr. Rodney Cocking (NSF) said in Granada: "Why OECD? Because I think we really need to have the international focus." (See the Granada report on the OECD website: *www.oecd.org/pdf/*M00017000/M00017849.*pdf*).

and connection. On the one hand, advancement of learning has depended, at least since the time of Aristotle, on the differentiation and specialisation of distinct fields of enquiry, faculties, disciplines or subjects: arts and science; the physical, biological and social sciences; physics, chemistry, biology, botany and zoology; and so on.[3] On the other, some of the most noteworthy advances have been achieved through the transference of insights across the boundaries of disciplines, or even the blurring of boundaries to create a new discipline.

In a recent paper addressing this issue, Dr. Koizumi[4] introduces the concept of trans-disciplinary studies to distinguish the creation of new sciences with their own conceptual structure from the mutual influence that typically occurs when established disciplines are contiguous or begin to overlap:

"Over the past two centuries, human culture has been split into two categories, science and technology, and the humanities and arts. Furthermore, science and technology have been minutely divided into clearly specified disciplines. Thus, it has become hard to understand other disciplines at a professional level because of the intellectual walls between disciplines. The maturity of science and technology, however, has made it increasingly difficult to obtain new findings and breakthroughs only within one's specialised discipline. New findings and technical breakthroughs are often accomplished only by bridging the gap between completely different disciplines, and this has been true for many years. For example, Newton's system of classic dynamics was created by combining the concepts

3. Since the beginning of Western science, basic scientific disciplines like physics, chemistry and biology have pursued with great success a reductionist methodological approach. This has led to most of our present understanding of the world around us and the functioning of our own body. Will such a reductionist approach help us to understand the complexity of the brain? Many scientists would argue that the brain is the most complex system in the universe (see "Beyond Reductionism", in *Science*, Vol. 284, April 1999). For instance, while software and hardware can be easily separated in a computer system, such an analogy, if applied to the brain, tells us that they are totally interwoven forming an extremely complex system. As pointed out by Christof Koch and Gilles Laurant (see "Complexity and the nervous system", in *Science*, Vol. 284, April 1999), the most obvious thing to say about the brain from a "complex system" point of view is that continued reductionism and atomisation will probably not on its own lead to fundamental understanding of the complexity of the brain. For instance, what possible links exist between the complexity of the nervous system and phenomena like consciousness and subjective experiences? It is therefore likely that the great scientific tradition of reductionism will have to be complemented with other scientific approaches trying to understand the "complex system" (possibly even including rather unexpected contributions, such as insights from quantum theory), in order to better understand the brain.
4. Basing his theory on the trans-disciplinarity that he observed with Pierre and Marie Curie: "(...) [they] benefited from the trans-disciplinarity of their thinking. (...) Trans-disciplinarity emerges through a fusion and bridging of disciplines, but not merely through the bundling of multiple disciplines." (Koizumi, H., *op.cit.*, pp. 9 and 19).

explaining the motion of astronomical objects and the falling of an object, traditionally said to be an apple, to the ground. Darwin's theory of natural selection was an analogy of the competition in a free market described by Adam Smith. (...) Although many scientists and scholars have recognised the importance of a multidisciplinary approach, it is still very difficult to transcend the borders of disciplines in practice. Such conceptual transitions have generally been made by people now considered geniuses. Current inter- or multi-disciplinary research organisations are not powerful enough to overcome the walls between disciplines, and inter- or multi-disciplinary research organisations often have not functioned as well as expected because they have been based upon only a bundle of closely, or sometimes not so closely, related disciplines. (...) Rather than a static concept, a dynamic concept is needed to overcome this difficulty (...)."

Figures *4a* and *4b* show the difference between the concepts of inter- or multidisciplinarity and trans-disciplinarity.

"New comprehensive fields, such as mind-brain science, environmental science as well as educational science, cannot be looked upon as a mere bundle or a simple combination of many related disciplines. Such fields apply the essence of knowledge and philosophy taken from many related disciplines to form their own conceptual structure, a structure that may transcend the borders of many natural sciences, social sciences, and even the humanities. The concepts of inter-disciplinarity or multi-disciplinarity are situated on a two-dimensional plane, but the trans-disciplinary concept occupies a three-dimensional space as shown in Figure *4b*. The trans-disciplinary concept exists at a higher hierarchical level produced by

Figure 4*a*. **Inter- and multi-disciplinarity**

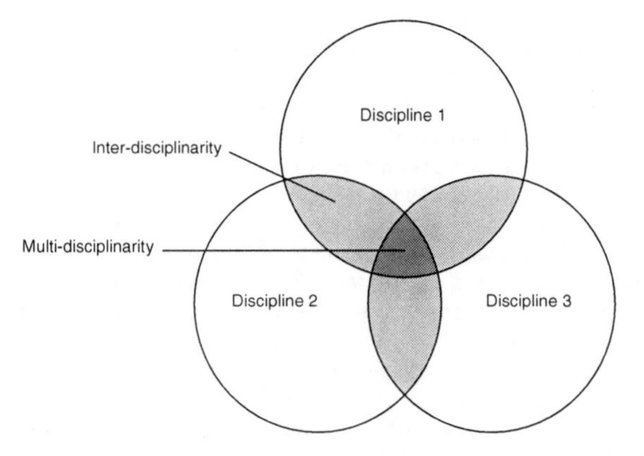

Figure 4*b.* **Trans-disciplinarity**

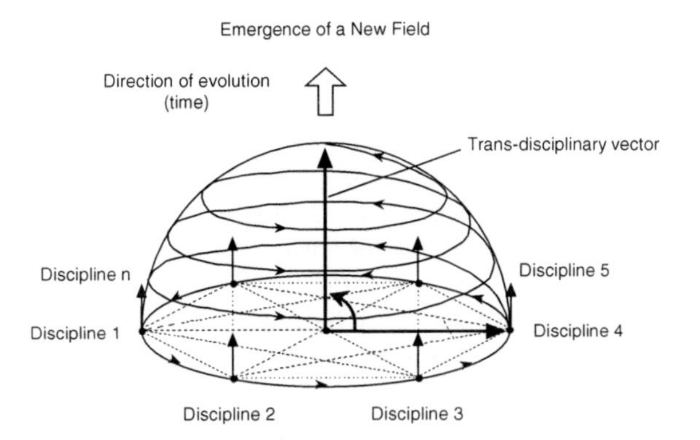

Source: Hideaki Koizumi (1999), "A practical approach towards trans-disiciplinary studies for the 21st century", *J. Seizon and Life Sci.*, Vol. 9, pp. 5-24.

the linkage of several different disciplines at the lower hierarchical level. Trans-disciplinarity includes the concept of bridging and fusion between completely different disciplines."[5]

Dr. Koizumi demonstrates that trans-disciplinary development does not happen easily or by chance.[6] The world needs an outstanding scholar, a Lucretius, a Darwin, a Terada[7] or a Teilhard de Chardin to establish the bridge or fusion between antecedent disciplines. In modern societies, it is important to review the structure of universities, the organisation of research, and indeed

5. Koizumi, H., *op. cit.*, pp. 6-8.
6. "Each discipline evolves by itself in terms of a conventional methodology and research organisations. However, some driving force is needed to bridge and fuse disciplines (...). A new comprehensive discipline will require new methodologies and new research organisations." (Koizumi, H., *op. cit.*, p. 8). As Dr. Eric Hamilton, Programme Director at NSF, underlined during the Tokyo forum: "(...) [in] our ROLE programme, the intent for building the structured inter-disciplinarity is really to look at the frontiers and the boundaries [of various disciplines], to look at problems that might be examined in one domain, but have answers that fall in other domains. (...) this is very consistent with the line of inquiry that was suggested [by Dr. Koizumi]: the trans-disciplinary vector".
7. Tarahiko Terada, Japanese physicist and essayist (1878-1935). As a professor of physics at the Tokyo Teikoku University, he produced 300 physics papers in his life, but is primarily remembered for his literary essays. Among other things, he fused the spirit of haikai poetry with the perspective of physics.

the school curriculum, to foster and encourage appropriate trans-disciplinary developments and provide a balance to the inevitably atomistic nature of natural science.

There can be no doubt of the value of trans-disciplinary relations in the field of "learning and the brain".[8] Previous chapters of this book provide numerous examples of fruitful insights arising from the experiment in trans-disciplinarity at the three fora.[9] Trans-disciplinarity is not a simple matter or a soft option. It must be both promoted and rewarded. Phase 2 of the OECD-CERI project offers a replicable model for the promotion of trans-disciplinary research. But even this bold initiative is a short-term measure. What is required is a determined and enduring commitment to trans-disciplinary relationships in the field of "learning and the brain" among the OECD Member countries.

The publication of this book both recognises and signals the emergence of a new science. The sciences of learning, including cognitive neuroscience, the cognitive sciences, medicine and education, are generally moving through inter-disciplinarity to trans-disciplinarity and transforming themselves into a new science of learning. This process is, as yet, at an early stage. But it is already clear that this transformation is both desirable and inevitable. Even at this stage it is by no means too soon to be considering the establishment of faculties and institutions of learning sciences either within existing universities and research centres, or free-standing and independent. There can be few questions more important, for the 21st century to find good answers to, than: how the brain works, how people learn best, and what educational provision can best help them. It will be the business of the science of learning to provide reliable and applicable answers to such questions. There are good reasons to believe that it will do so in the years ahead.

8. "Some researchers at the frontier of mind-brain science have begun to realise the value of a trans-disciplinary approach. These researchers are now studying, for example, philosophy, psychology, linguistics even though their original fields might be mathematics, physics, chemistry, physiology or medicine. Some of them are now almost ready to employ the trans-disciplinary approach to scientifically study 'consciousness'" (Koizumi, H., *op. cit.*, p. 19).
9. "Why OECD? (...) To help us move along and bring our fields of science together. OECD can now help us move the learning sciences into an evidence based science; it is really important to bring together the trans-disciplinary approaches, the different perspectives, the different methodologies, the different technologies, because [in order] to move to the evidence-based sciences of learning, we need to be also looking at the convergence of the evidence of all these fields" (Dr. Rodney Cocking at the Granada forum).

5.2. The next steps: research networks

"Research is always incomplete."
Mark Pattison

5.2.1. *Research types and methodology*

There are four types of research, of which at least three are pertinent to meeting the outcomes of the second phase of the "Learning Sciences and Brain Research" project. The first one concerns synthesising and collecting data pertaining to existing research from many different sources (universities, research institutions, etc.). The second type of research involves the continuation of on-going work (projects and programmes that are currently in progress). The third type is extension research, in which recently finished work needs a related extension in order to further test and refine hypotheses. The fourth type is new research and includes projects engaged in related hypotheses, new theory proposal, and testing. Phase 2 of the project will mostly include the first three types of research, but also, to some extent, the fourth one, as needed.

5.2.2. *Three research areas*

Area 1: Brain development and literacy

A considerable number of children and adults world-wide still exhibit difficulty in learning to read, spell and write words; some even suffer from various forms of dyslexia. Research through brain imaging technologies has made substantial progress in understanding how literacy develops in the brain. Additionally, important neuroscientific research findings are emerging that show how deficiencies affect the brain. Understanding them can better aid educators in reacting appropriately to these disabilities.

Research objectives include:

- understanding how the brain engages in reading, including visual processes with a focus on how different brain areas work together;
- visual and attentional studies of dyslexia, including interventions and strategies for amelioration;
- language learning and the effects of brain plasticity in both adults and children along with a focus on interventions linked to brain imaging.

Area 2: Brain development and numeracy

The lack of mathematical competence continues to cause countless children difficulty in school with later repercussions in adulthood (including dyscalculia).

An important body of new brain research is helping educators to understand ways in which a child's number sense develops from a very early age. This area will also focus on symbolic thinking and on how the computer can be used to both diagnose and ameliorate deficiencies in mathematics learning. Current research has begun to show that early intervention can be facilitated through appropriate ICT intervention and rehabilitation tools.

Research objectives include:

- developmental trajectory of numeracy including mental models for early math (including the establishment of a timeline for math skills acquisition and studies in gender differences);
- taxonomy of dyscalculias including relations with dyslexia, diagnostic tests, biological *versus* social origins, tools for learning and remedial strategies;
- evaluating and designing diagnosis and intervention strategies, including school strategies and training studies in adults;
- ICT tools for mathematics learning and acquisition of ICT literacy.

Area 3: Brain development and learning over the life cycle

One of the most important insights emanating from Phase I of this project is the strong agreement in the scientific community about the brain plasticity over the life-span. Research has shown that plasticity, the capacity to learn, unlearn and relearn over the life cycle, is much greater than was previously recognised. At the same time, the brain's "sensitive periods" require further attention. This area of research not only addresses individual learning styles, but is also of fundamental importance to education policies for lifelong learning, and for the care of the ageing brain. Research work is exploring ways in which normal age-related decline, as well as disease-related deterioration, can be slowed. How to shift this work from the laboratory to formal and informal learning activities of adults is a key issue for education policy and practice.

Research objectives include:

- age-related capacities as they relate to cognition and perception, including interventions and strategies;
- capacities in infancy and early childhood, with attention to sensitive periods, stress/maternal influence and speech sound learning in babies;
- cognitive mechanisms of learning, in childhood and adolescence focusing on cognitive control of emotions, development of sensorimotor functions, learning of art and music and formation of the self;
- cognitive mechanisms of learning in adults and ageing people, including functional reorganisation in aged damaged brains, how to learn in an

information-based aged society, and delaying the declines due to the ageing process;

- cognitive mechanisms of learning, including individual differences, with respect to age and implicit and explicit learning.

5.2.3. *Three research networks: structure and expected outcomes*

Each of the three research areas detailed above will be addressed over the three-year period by an international and trans-disciplinary network, composed mainly of neuroscientists and education experts. Each of the three networks is led by a Main Co-ordinator and a Network Advisor. These six persons compose the project's "Steering Committee". Their task will be to provide co-ordination within and ensure cross-fertilisation between the networks. The LLL Network (for "Life-long Learning") will be co-ordinated from Asia.[10] The LRS Network (for "Literacy and Reading Skills") will be co-ordinated from America.[11] The NMS Network (for "Numeracy and Maths Skills") will be co-ordinated from Europe.

Moreover, a "Scientific Advisory Group" composed of external experts, along with one representative of each network, will follow the activities of the networks and provide advice and recommendations on substantive and operational orientations. These two bodies (the Steering Committee and the Scientific Advisory Group) will be continually assisted by staff from OECD-CERI.

Apart from reviewing and codifying existing research findings, co-ordinating and facilitating ongoing research and (where appropriate) encouraging and stimulating relevant new research, the networks will be required to produce regular progress reports, contribute insights and provide substantial chapters in an OECD publication (scheduled for 2005), including reports on "the state of the science" findings and conclusions to date, recommendations for policy and practice, and advice on the research agenda for the next phase.

The focus and need for research on life-long learning emanates from both societal and biological pressures and demands. With societies struggling with their ageing populations, there is a turning toward neuroscience to understand what is appropriate for age groups and when. Additionally, with rapid globalisation, there is a need for human brains to adapt to and assimilate changes. Hence, implementation of the research objectives for the LLL Network will follow a framework of firstly systematising a knowledge base specific to each research objective

10. Co-ordinator: Pr. Masao Ito (RIKEN Brain Science Institute, Tokyo, Japan); Network Advisor: Dr. Takao Hensch (idem).
11. Co-ordinator: Pr. Michael Posner (Sackler Institute, New York City, USA); Network Advisor: Dr. Bruce McCandliss (idem).

(including the definition of key concepts; for example, sensitive *versus* reliable periods for early childhood learning) and secondly opening communication mediums to better address what educators and policy-makers want and need to know (through FAQ's on a website, for example). Overarching issues for research into life-long learning include brain function as it pertains to: nutrition, sleep, drug consumption, stress management, physical fitness, and emotional regulation. This network seeks to responsibly incorporate basic hard science into policy recommendations and/or directions for policy-makers, educators, healthcare providers and parents.

The overarching goals seen for the LRS Network are dissemination and innovative application. Target audiences are both education policy-makers and end users, which include: educationalists, parents and children. The general guidelines to be followed to ensure network focus are the following: 1) research must be grounded in brain mechanisms and 2) there must be a utility and accessibility of information. The international context of this endeavour should be a horizontal issue present in each research domain. As such, this network proposes a strong ICT component and would use a website to create a framework that focuses international readers on established points of agreement in the brain literature as it applies to literacy, and to link to discussions where world scientists can state evidence for or against a particular topic, including discussion around "neuromythologies". Interactive web content will also include demonstrations and keys to intervention tools.

Overall goals in the NMS Network for all research domains listed include 1) the identification of key cerebral structures and their interconnections, 2) the determination of the interaction between mathematics and space, and 3) understanding the impact of cultural differences and interactions. Each of the three research domains will provide a synthesis of scientific knowledge, a list of tentative educational consequences, a list of open scientific and educational questions including new research proposals. Additionally, this network proposes to work horizontally with the LLL Network to more completely understand how learning mathematics relates to learning in other domains and with the LRS Network to research the connections between dyslexia and dyscalculia. This network proposes a strong ICT component by introducing new diagnostic and intervention literacy software games for children and adults freely available on the Internet.

Although neuroscientific research into literacy, mathematical thought, and learning over the life-span is underway in several scientific institutions, most are too scattered to span the distance from laboratory to learning systems. With these international teams of trans-disciplinary scientists, research into the above listed areas will be co-ordinated on a collaborative basis. A principal objective is to collect and publish useful research that can directly impact educational practice.

A second objective is to disseminate this information, through both print and multimedia, to targeted audiences including educational practitioners and policy-makers. This concept includes a bridging and fusing of different disciplines, among which linguistics, mathematics, psychology, and of course, cognitive neuroscience and philosophy.

Annex

AGENDAS OF THE THREE FORA

Brain Mechanisms and Early Learning

First High Level Forum, 16-17 June 2000, Sackler Institute, New York City, USA

Opening

Michael Posner, Director of Sackler Institute, USA

Jarl Bengtsson, Head of CERI/OECD

Session 1: Synthesis of Brain Research and Learning Sciences

Michael Posner, Sackler Institute, USA
"Linking brain development to education"

Andries Sanders, Vrije Universiteit Amsterdam, The Netherlands
"Potential relevance of brain research to learning processes and educational curricula for pre-school children"

William Greenough, University of Illinois, USA
"Brain's mechanisms of learning and memory"

Session 2: Cognition and Emotion

Joseph LeDoux, New York University, USA
"Personality and the brain: closing the gap"

Masao Ito, RIKEN-Brain Science Institute, Japan
"Brain mechanisms of cognition and emotion"

David Servan-Schreiber, University of Pittsburgh, USA
"Emotional context of learning"

Stephen Kosslyn, Harvard University, USA
"Using mental imagery to regulate emotion"

Session 3: Numeracy, Literacy and Language Acquisition

Stanislas Dehaene, INSERM, France
"Brain mechanisms of numeracy"

Helen Neville, University of Oregon, USA
"Brain mechanisms of first and second language acquisition"

Bruce McCandliss, Sackler Institute, USA
"Cortical circuitry of word reading"

Session 4: Learning and The Brain – Relevance of Interdisciplinary Approaches

Rodney Cocking, National Science Foundation, USA
"New developments in the science of learning: using research to help students learn"

Alison Gopnik, University of California, Berkeley, USA
"Cognitive development and learning sciences: state of the art"

Concluding Session

Michael Posner, Sackler Institute, USA
"Scientific reflections on the results of the forum"

Eric Hamilton, National Science Foundation, USA
"Policy reflections on the results of the forum"

Sir Christopher Ball, University of Derby, UK
"General conclusions of the forum"

Brain Mechanisms and Youth Learning

Second High Level Forum, 1-3 February 2001, University of Granada, Spain

Opening

José Moratalla Molina, Mayor of Granada

David Aguilar Peña, Rector of the University of Granada

Jarl Bengtsson, Head of CERI/OECD

Session 1: Synthesis of Previous Forum and Outlines

Bruno della Chiesa, CERI/OECD
"Synthesis of the main results of the first forum"

John Bruer, McDonnell Foundation, USA
"Brain science, mind science, and learning across the life-span"

Pilar Ballarin, Region Andalusia, Spain
"Main policy questions at high school level"

Session 2: Bridging Neurosciences and Genetic Issues

Luis Fuentes, University of Almería, Spain
"Bridging neurosciences and the genome research"

Antonio Marín, University of Seville, Spain
"Genetics and intellectual performance"

Stanislas Dehaene, INSERM, France
"Impact of early brain damage on childhood acquisition of mathematics"
(fetal alcoholism syndrome, Turner syndrome and other genetic diseases)

Jim Swanson, University of California, Irvine, USA
"Genetic factors associated with ADHD"

Rafael Maldonado, University Pompeu Fabra, Barcelone, Spain
"Impact of drug consumption on Learning"

Alain Michel, Ministry of Education, France
"Ethical issues related to genetics and neuroscience: educational perspectives"

Session 3: Adolescents' Context of Learning and Learning Modes

Pio Tudela, University of Granada, Spain
"Implicit and explicit learning: a cognitive neuroscience point of view"

José Manuel Rodriguez-Ferrer, University of Granada, Spain
"Specificities of post-puberty period: harnessing hormones?"

Stephen Kosslyn, Harvard University, USA
"The role of mental simulation in thinking"

Heinz Schirp, Institute for School and Continuing Education, Germany
"Adolescents' learning, from an educational policy point of view"

Session 4: Numeracy, Literacy and Creativity

Diego Alonso, University of Almería, Spain
"Brain mechanisms of youth acquiring mathematic skills"

Bruce McCandliss, Sackler Institute, USA
"Brain mechanisms of reading skills: from novice to expert"

Guy Claxton, University of Bristol, UK
"How brains make creativity, and how schools strengthen or weaken young people's creative birthright"

Mark Fletcher, English Experience, UK
"A classroom challenge to neuroscience (and to education):
The brain-friendly revolution – Reality or neuro-babble?"

Rodney Cocking, National Science Foundation, USA
"Reflections on session 4"

Concluding Session

Pilar Ballarin, Region Andalusia, Spain
"Policy reflections on the results of the forum"

Masao Ito, RIKEN-Brain Science Institute, Japan
"Scientific reflections on the results of the forum and opening to Tokyo"

Jean-Claude Ruano-Borbalan, "Sciences Humaines", France
"Reflections on the forum, from a scientific journalist's point of view"

Sir Christopher Ball, University of Derby, UK
"General conclusions of the forum"

Concluding Words

José Moratalla Molina, Mayor of Granada, Spain

Julio Iglesias de Ussell, Secretary of State for Universities, Ministry of Education, Spain

Candida Martinez, Regional Minister for Education/Science, Region Andalusia, Spain

Jarl Bengtsson, Head of CERI/OECD

Bruno della Chiesa, Administrator, CERI/OECD

Brain Mechanisms and Learning in Ageing

Third High Level Forum, 26-27 April 2001, RIKEN-Brain Science Institute, Tokyo, Japan

Opening

Teiichi Sato, Director General – JSPS, Japan

Masao Ito, Professor and Director of RIKEN-Brain Science Institute, Japan

Jarl Bengtsson, Head of CERI/OECD

Session 1: Introduction and Scientific Overview

Eamonn Kelly, George Mason University, Fairfax, USA
"Synthesis of the main results of the two first fora"

Raja Parasuraman, Catholic University of America, Washington DC, USA
"Attention, ageing, and dementia: extending and enhancing cognitive function in adulthood"

Session 2: Interdisciplinary Issues related to Learning in Ageing

Jarl Bengtsson, CERI/OECD
"Ageing populations: New policy challenges"

Shinobu Kitayama, University of Kyoto, Japan
"Cultural variations in cognition: implications for ageing research"

Yasumasa Arai, Juntendo University, Tokyo, Japan
"Gender issues: is there a sexual brain?"

Rodney Cocking , National Science Foundation, USA
"Crossing disciplinary boundaries to understand the cognitive neuroscience of ageing"

Hideaki Koizumi, Advanced Research Laboratory, Hitachi Ltd., Japan
"Reflections on sessions 1 and 2"

Session 3: Brain Plasticity over the Life Cycle, Memory and Lifelong Learning

Andrea Volfova, Harvard University, USA
Bruno della Chiesa, CERI/OECD
"What could brain plasticity mean for lifelong learning?"

Yasushi Miyashita, University of Tokyo, Japan
"Memory: encoding and retrieval"

Itaru Tatsumi, Tokyo Metropolitan Institute of Gerontology, Japan
"A PET activation study on retrieval of proper and common nouns in young and elderly people"

Lynn Cooper, University of Columbia, USA
"Age-related effects on dynamic properties of dissociable memory systems"

Session 4: Skills Acquisition, Later in Life

Masao Ito, RIKEN-Brain Science Institute, Japan
"Roles of the cerebellum in skills acquisition and its dependence of age"

Pio Tudela, University of Granada, Spain
"Cognitive skills acquisition, later in life: attention and automaticity"

Wolfgang Schinagl, IHK Steiermark, Austria
"New learning of adults in the information and knowledge society"

Bruce McCandliss, Sackler Institute, New York, USA
"Brain mechanisms influencing adult learning:
the case of persistent difficulties in learning non-native speech sounds"

Kenneth Whang, National Science Foundation, USA
"Reflections on sessions 3 and 4"

Session 5: Diseases, Learning and the Power of the Ageing Brain

Shigenobu Kanba, Yamanashi Medical School, Japan
"Characteristics of senile depression: importance of prevention and treatment"

Akihiko Takashima, RIKEN-Brain Science Institute, Japan
"Understanding the ageing brain from studies of Alzheimer's disease"

Art Kramer, University of Illinois, USA
"Enhancing the cognitive vitality of older adults: the role of fitness and cognitive training"

Yoshiko Shimonaka, Bunkyo Women's University, Japan
"Creativity and ageing: does creativity decline in the adult life-span?"

Session 6: Learning and Education: Research Policy Perspectives

Akito Arima, Former Minister of Education and Science, Japan
"Education and research in Japan"

Denis Ralph, South Australian Centre for Lifelong Learning and Development, Australia
"Learning across the life-span – Linking research, policy and practice: an Australian perspective"

Eric Hamilton, National Science Foundation, USA
"NSF policy and programmes on brain research and learning sciences"

Barry McGaw, DEELSA/OECD
"Reflections on sessions 5 and 6"

Concluding Session

Masao Ito, RIKEN-Brain Science Institute, Japan
"Scientific reflections on the work"

Sir Christopher Ball, University of Derby, UK
"Policy reflections on the work and general conclusions of the Phase I fora"

Bruno della Chiesa, CERI/OECD
"Next Steps: towards Phase 2"

References

I. Books

A. *Introductory texts*

Bruer, J.T. (1999),
The Myth of the First Three Years: A New Understanding of Early Brain Development and Lifelong Learning, Free Press, New York.

Carter, R. (2000),
Mapping the Mind, University of California Press, Berkeley, CA.

Dehaene, S. (1997),
The Number Sense: How the Mind Creates Mathematics, Oxford University Press, New York.

Goleman, D. (1995),
Emotional Intelligence, Bantam Books, New York.

Gopnik, A., Meltzolf, A. and Kuhn, P. (1999),
The Scientist in the Crib, William Morrow and Co., New York.

Ito, M. (1997),
Brain and Mind, Elsevier Science, UK.

Kosslyn, S.M. (1996),
Image and Brain, MIT Press, Cambridge, MA.

National Research Council (1999),
How People Learn: Brain, Mind, Experience and School, National Academy Press, Washington DC.

Parasuraman, R. (1998),
The Attentive Brain, MIT Press, Cambridge, MA.

Pinker, S. (2000),
The Language Instinct: How the Mind Creates Language, Harper Perenniel, San Francisco, CA.

Posner, M. I. and Raichle, M. (1994),
Images of Mind, Scientific American Books.

Spitzer, M. (1999),
The Mind Within the Net: Models of Learning, Thinking, and Acting, MIT Press, Cambridge, MA.

B. *For further reading*

Ball, C. (1989),
Higher Education into the 1990's: New Dimensions, Open University Press, UK.

Ball, C. (1991),
Learning Pays, RSA, London.

Byrnes, J.P. (2001),
Minds, Brains and Learning: Understanding the Psychological and Educational Relevance of Neuroscientific Research, The Guilford Press, New York.

Claxton, G. (1999),
Wise-Up: The Challenge of Lifelong Learning, Bloomsbury Publishing, UK.

Damasio, A.R. (1994),
Descartes' Error: Emotion, Reason and the Human Brain, Putnam, New York.

Damasio, A.R. (1999),
The Scientific American Book of Brain, The Lyons Press, New York.

Foder, J.A. (1983),
The Modularity of Mind, MIT Press, Cambridge, MA.

Gardner, H. (1983),
Frames of Mind, London.

Gazzaniga, M.S. (1996),
Conversations in the Cognitive Neurosciences, MIT Press, Cambridge, MA.

Hebb, D.O. (1949),
The Organization of Behavior: A Neuropsychological Thoery, Wiley Publishing, New York.

Ito, M. (1967),
The Cerebellum as a Neuronal Machine, Springer-Verlag, UK.

Ito, M. (1984),
The Cerebellum and Neural Control, Raven Press, UK.

Kotulak, R. (1997),
Inside the Brain: Revolutionary Discoveries of How the Mind Works, Andrews McMeel Publishing, Kansas City, KS.

OECD (2000),
Knowledge Management in the Learning Society, Paris.

OECD (2000),
Learning to Bridge the Digital Divide, Paris.

OECD (2001),
Cities and Regions in the New Learning Economy, Paris.

OECD (2001),
E-Learning – The Partnership Challenge, Paris.

OECD (2001),
Knowledge and Skills for Life – First Results from PISA 2000, Paris.

OECD (2001),
Learning to Change: ICT in Schools, Paris.

OECD (2001),
New York report: *www.oecd.org/pdf/M00019000/M00019809.pdf*

OECD (2001),
Granada report: *www.oecd.org/pdf/M00017000/M00017849.pdf*

OECD (2001),
Tokyo report: *www.oecd.org/pdf/M00022000/M00022657.pdf*

OECD (2001),
Reports of the three fora: *www.oecd.org/oecd/pages/home/displaygeneral/0,3380,EN-document-603-5-no-27-26268-0,FF.html*

Spitzer, M. (2000),
Geist, Gehirn and Nervenheilkunde: Grenzgänge zwischen Neurobiologie, Psychopathologie und Gesellschaft, F.K. Schattauer Verlagsgesellschaft mbH, Stuttgart.

Spitzer, M. (2001),
Ketchup und das kollektive Unbewusste: Geschichten aus der Nervenheilkunde, F.K. Schattauer Verlagsgesellschaft mbH, Stuttgart.

Thompson, W.L. and Kosslyn, S.M. (in press),
"Neuronal systems activated during visual mental imagery: A review and meta-analyses", in J. Mazziotta and A. Toga (eds.), *Brain Mapping II: The Systems*, Academic Press, New York.

US National Research Council,
Report on Scientific Inquiry in Education.

II. Articles

A. *Mentioned in this book and/or in the three fora reports*

Albert, M.S., Jones, K., Savage, C.R., Berkman, L., Seeman, T., Blazer, D. and Rowe, J.W. (1995),
"Predictors of cognitive change in older persons: MacAuthur studies of successful ageing", *Psychological Ageing*, Vol. 10, No. 4, pp. 578-589.

Alexopoulos, G.S., Meyers, B.S., Young, R.C., Kakuma, T., Silbersweig, D. and Charlson, M. (1997),
"Clinically defined vascular depression", *American Journal of Psychiatry*, Vol. 154, pp. 562-565.

Block, J. (1995),
"On the relation between IQ, impulsivity and delinquency", *Journal of Abnormal Psychology*, Vol. 104, pp. 395-398.

Braak, H. and Braak, E. (1991),
"Neuropathological stageing of Alzheimer-related changes", *Acta Neuropathologica*, Vol. 82, pp. 239-259.

Bruer, J.T. (1998),
"Brain science, brain fiction", *Educational Leadership*, Vol. 56, No. 3, pp. 14-18.

Bruer, J.T. (1999),
"Education and the brain: A bridge too far", *Educational Researcher*, Vol. 26, No. 8, pp. 4-16.

Bruer, J.T. (1999),
"In search of brain-based education", *Phi Delta Kappan*, Vol. 80, No. 9, pp. 648-657.

Burgess, N. and O'Keefe, J. (1996),
"Neural computation underlying the firing of place cells and their role in navigation", *Hippocampus*, Vol. 6, No. 6, pp. 749-762.

Bush, G., Luu, P. and Posner, M.I. (2000),
"Cognitive and emotional influences in anterior cingulate cortex", *Trends in Cognitive Neuroscience*, Vol. 4, No. 6, pp. 215-222.

Dehaene, S., Spelke, E., Pinel, P., Stanescu, R. and Tsivlin, S. (1999),
"Sources of mathematical hinking: Behavioural and brain imaging evidence", *Science*, Vol. 284, No. 5416, pp. 970-974.

Diamond, M.C., Greer, E.R., York, A., Lewis, D., Barton, T. and Lin, J. (1987), "Rat cortical morphology following crowded-enriched living conditions", *Experimental Neurology*, Vol. 96, No. 2, pp. 241-247.

Dustman, R.E., Shearer, D.E. and Emmerson, R.Y. (1993), "EEG and event-related potentials in normal ageing", *Progressive Neurobiology*, Vol. 41, No. 3, pp. 369-401.

Ernst, R.L. and Hays, J.W. (1994), "The US economic and scocial costs of Alzheimer's disease revisited", *American Journal of Public Health*, Vol. 84, pp. 1261-1264.

Eslinger, P.J. and Damasio, A.R. (1985), "Severe disturbance of higher cognition after bilateral frontal lobe ablation: Patient EVR", *Neurology*, Vol. 35, pp. 1731-1741.

Felsman, J.K. and Vaillant, G.E. (1987), "Resilient children as adults: A 40-year study", in E.J. Anderson and B.J. Cohler (eds.), *The Invulnerable Child*, Guilford Press, New York.

Gabrieli, J.D., Brewer, J.B. and Poldrack, R.A. (1998), "Images of medial temporal lobe functions in human learning and memory", *Neurobiology of Learning and Memory*, Vol. 20, No. 1-2, pp. 275-283.

Goldsmith, H.H. and Bihun, J.T. (1997), "Conceptualizing genetic influences on early behavioral development", *Acta Paediatric*, July, Vol. 422, pp. 54-59.

Greenwood, P.M., Sunderland, T., Friz, J. and Parasuraman, R. (2000), "Genetics and visual attention: Selective deficits in healthy adult carriers of the e4 allele of the apolipoprotein E gene", Proceedings of the National Academy of Sciences, Unites States, Vol. 97, pp. 11661-11666.

Hubel, D.H., Wiesel, T.N. and LeVay, S. (1977), "Plasticity of ocular dominance columns in monkey striate cortex", *Philosophical Transactions of the Royal Society of London* (B), Vol. 278, pp. 307-409.

Koizumi, H. (1997), "Mind-morphology: an approach with non-invasive higher-order brain function analysis", *Chemistry and Chemical Industry*, Vol. 50, No. 11, pp. 1649-1652.

Koizumi, H. (1999), "A practical approach to trans-disciplinary studies for the 21st century – The centennial of the discovery of radium by the Curies", *J. Seizon and Life Sci.*, Vol. 9, No. B 1999.1, pp. 19-20.

Koizumi, H. *et al.* (1999), "Higher-order brain function analysis by trans-cranial dynamic near-infrared spectroscopy imaging", *Journal Biomed. Opt.*, Vol. 4, front cover and pp. 403-413.

Maguire, E.A., Frackowiak, R.S. and Frith, C.D. (1996), "Learning to find your way around: A role for the human hippocampal formation", *Proceedings for the Royal Society of London* (B): *Biological Sciences*, Vol. 263, pp. 1745-1750.

Maguire, E.A., Frackowiak, R.S. and Frith, C.D. (1997), "Recalling routes around London: Activation of the right hippocampus in taxi drivers", *Journal of Neuroscience*, Vol. 17, No. 18, pp. 7103-7110.

Maguire, E.A., Gadian, D.S., Johnsrude, I.S., Good, C.D., Ashburner, J., Frackowiak, R.S. and Frith, C.D. (2000),
"Navigation related structural changes in the hippocampi of taxi drivers", Proceedings of the National Academy of Sciences, United States, Vol. 97, No. 8, pp. 4398-4403.

McCandliss, B.D., Beck, I., Sandak, R. and Perfetti, C. (in press),
"Focusing attention in decoding for children with poor reading skills: A study of the Word Building intervention".

Meltzoff, A.N. and Moore, M.K. (1977),
"Imitation of facial and manual gestures by human neonates", *Science*, Vol. 198, pp. 4312, pp. 75-78.

O'Connor, T.G., Bredenkamp, D. and Rutter, M. (1999),
"Attachment disturbances and disorders in children exposed to early severe deprivation", *Infant Mental Health Journal*, Vol. 20, No. 10, pp. 10-29.

Pantev, C., Osstendveld, R., Engelien, A., Ross, L.E., Roberts, L.E. and Hoke, M. (1998),
"Increased auditory cortical representation in musicians", *Nature*, Vol. 392, pp. 811-814.

Parasuraman, R. and Greenwood, P.M. (1998),
"Selective attention in aging and dementia", in R. Parasuraman (ed.), *The Attentive Brain*, pp. 461-488, MIT Press, Cambridge, MA.

Parasuraman, R. and Martin, A. (1994),
"Cognition in Alzheimer's disease: Disorders of attention and semantic knowledge", *Current Opinion in Neurobiology*, Vol. 4, pp. 237-244.

Park, D.C. (2001),
The Ageing Mind. See website: *www.rcgd.isr.umich.edu/*

Pascual-Leone, A., Nguyet, D., Cohen, L.G., Brasil-Neto, J.P., Cammarota, A. and Hallett, M. (1995),
"Modulation of muscle responses evoked by transcranial magnetic stimulation during the acquisition of new fine motor skills", *Journal of Neurophysiology*, Vol. 74, No. 3, pp. 1037-1045.

Raz, N., Williamson, A., Gunning-Dixon, F., Head, D. and Acher, J.D. (2000),
"Neuroanatomical and cognitive correlates of adult age differences in acquisition of a perceptual-motor skill", *Microscience Research Technology*, Vol. 51, No. 1, pp. 85-93.

Rothbart, M.K. and Jones, L.B. (1998),
"Temperament, self-regulation and education", *School Psychology Review*, Vol. 27, No. 4, 479-491.

Schinagl, W. (2001),
"New learning of adults in the information and knowledge society", *Journal of Universal Computer Science*, Vol. 7, No. 7, pp. 623-628.

Shaywitz, S.E., Shaywitz, B.A., Pugh, K.R., Fulbright, R.K., Constable, R.T., Mencl, W.E., Shankweiker, D.P., Liberman, A.M., Skudlarski, P., Fletcher, J.M., Katz, L., Marchione, K.E., Lacadie, C., Gatenby, C. and Gore, J.C. (1998),
"Functional disruption in the organisation of the brain for reading in dyslexia", Proceedings of the National Academy of Sciences, United States, Vol. 95, No. 5, pp. 2636-2641.

Shoda, Y., Mischel, W. and Peake, P.K. (1990),
"Predicting adolescent cognitive development and self-regulatory competencies from preschool delay of gratification: Identifying diagnostic conditions", *Developmental Psychology*, Vol. 26, pp. 978-986.

Terry, R.D., DeTeresa, R. and Hansen, L.A. (1987),
"Neocortical cell counts in normal human adult ageing", *Annuals of Neurology*, Vol. 21, No. 6, pp. 530-539.

B. *For further reading*

Hickok, G. and Poeppel, D. (2000),
"Towards a functional neuroanatomy of speech perception", *Trends in Cognitive Science*, Vol. 4, No. 4, pp. 131-138.

Huttenlocker, P.R. and Dabholkar, A.S. (1997),
"Regional differences in synaptogenesis in human cerebral cortex", *Journal of Computational Neurology*, Vol. 387, No. 2, pp. 167-178.

Kuhl, P.K. (1998),
"The development of speech and language", in T.J. Carew, R. Menzel and C.J. Shatz (eds.), *Mechanistic Relationships Between Development and Learning*, pp. 53-73, Wiley, New York.

Posner, M.I. and Abdullaev, Y. (1996),
"What to image? Anatomy, plasticity and circuitry of human brain function", in A.W. Toga and J. C. Mazziotta (eds.), *Brain Mapping: The Methods*, pp. 407-421, Academic Press, New York.

Posner, M.I., Abdullaev, Y.G., McCandliss, B.D. and Sereno, S.C. (1999),
"Neuroanatomy, circuitry and plasticity of word reading", *Neuroreport*, Vol. 10, pp. R12-23.

Stanescu-Cosson R., Pinel, P., van De Moortele, P. F., Le Bihan, D., Cohen, L. and Dehaene, S. (2000),
"Understanding dissociation's in dyscalculia: A brain imaging study of the impact of number size on the cerebral networks for exact and approximative calculation", *Brain*, Vol. 123, pt 11, pp. 2240-2255.

Temple, E., Poldrack, R.A., Salidis, J., Deutsch, G.K., Tallal, P., Mersenich, M.M. and Gabrieli, J.D. (2001),
"Disrupted neural responses to phonological and orthographic processing in dyslexic children: An fMRI study", *Neuroreport*, Vol. 12(2), pp. 299-307.

Glossary

Acalculia

See dyscaculia

ADHD

Attention Deficit Hyperactivity Disorder. A syndrome of learning and behavioural problems. Characterised by difficulty in sustaining attention, impulsive behaviour (as in speaking out of turn), and often by hyperactivity – also referred to as minimal brain dysfunction.

Alzheimer's disease

A progressive degenerative disease of the brain associated with ageing, characterised by diffuse atrophy throughout the brain with distinctive lesions called senile plaques and clumps of fibrils called neurofibrillary tangles. Cognitive processes of memory and attention are affected.

Amygdala

A part of the brain involved in emotions, emotional learning, and memory. Each hemisphere contains an amygdala, shaped like an almond and located deep in the brain, near the inner surface of each temporal lobe.

Angular gyrus

An area of the cortex in the parietal lobe associated with processing the sound structure of language and associated with reading.

Apolipoprotein E

(Or "apoE"). Has been studied for many years for its involvement in cardiovascular diseases. It has only recently been found that one allele (gene factor) of the apoE gene (E4) is a risk factor for Alzheimer's disease.

Cerebellum

A part of the brain located at the back and below the principal hemispheres, involved in the regulation of movement.

Choline

A chemical required for synthesising acetylcholine, a neurotransmitter needed for memory storage and muscle control.

Cholinergic systems

Also acetylcholine systems. Systems in which the neurotransmitter acetylcholine is present, which occur at neuromuscular junctions between motor neurons and the brain. The loss of acetylcholine (Ach) neurons is a contributing factor in Alzheimer's disease.

Cognition
Operation of the mind which includes all aspects of perceiving, thinking, learning, and remembering.

Cognitive neuroscience
Study and development of mind and brain research aimed at investigating the psychological, computational, and neuroscientific bases of cognition.

Cognitive science
Study of the mind. An interdisciplinary science that draws upon many fields including neuroscience, psychology, philosophy, computer science, artificial intelligence, and linguistics. The purpose of cognitive science is to develop models that help explain human cognition – perception, thinking, and learning.

Cognitive vitality
Refers to the active strength or force of mind throughout the life-span.

(cerebral) Cortex
Outer layer of the brain.

Decoding
An elementary process in learning to read alphabetic writing systems (for example, English, Spanish, German or Italian), in which unfamiliar words are deciphered by associating the letters within the word to corresponding speech sounds.

(senile) Dementia
A condition of deteriorated mentality that is characterised by marked decline from the individual's former intellectual level and often by emotional apathy. Alzheimer's disease is one form of dementia.

Depression
A lowering of vitality or functional activity: the state of being below normal in physical or mental vitality.

Dyscalculia
Impairment of the ability to perform simple arithmetical computations, despite conventional instruction, adequate intelligence and socio-cultutal opportunity.

Dyslexia
A disorder manifested by difficulty in learning to read despite conventional instruction, adequate intelligence, and socio-cultural opportunity.

EEG
Electroencephalogram. A measurement of the brain's electrical activity via electrodes. EEG is derived from sensors placed in various spots on the scalp, which are sensitive to the summed activity of populations of neurons in a particular region of the brain.

Emotional intelligence
Sometimes referred to as emotional quotient ("EQ"). Individuals with emotional intelligence are able to relate to others with compassion and empathy, have well-developed social skills, and use this emotional awareness to direct their actions and behaviour. The term was coined in 1990.

Epilepsy

A chronic nervous disorder in humans which produces convulsions of greater or lesser severity with clouding of consciousness; it involves changes in the state of consciousness and of motion due to either an inborn defect or a lesion of the brain produced by tumour, injury, toxic agents, or glandular disturbances.

ERP

Event-related potentials. Electrical signals are first recorded with an EEG. Data from this technology is then time locked to the repeated presentation of a stimulus to the subject, in order to see the brain in action. The resulting brain activation (or event-related potentials) can then be related to the stimulus event.

Experience-dependent

A property of a functional neural system in which variations in experience lead to variations in function, a property that might persist throughout the life-span.

Experience-expectant

A property of a functional neural system in which the development of the system has evolved to critically depend on stable environmental inputs that are roughly the same for all members of a species (*i.e.* stimulation of both eyes in new-borns during development of ocular dominance columns). This property is thought to operate early in life.

Explicit memory

Memories that can be retrieved by a conscious act, as in recall, and can be verbalised, in contrast to implicit or procedural memories, which are less verbally explicit.

Fetal alcohol syndrome

Sum of the damage done to the child before birth as a result of the mother drinking alcohol during pregnancy.

fMRI

Functional Magnetic Resonance Imaging. Use of an MRI scanner to view neural activity indirectly through changes in blood chemistry (such as the level of oxygen) and investigate increases in activity within brain areas that are associated with various forms of stimuli and mental tasks (see MRI).

Fragile X syndrome

One of the most common causes of inherited mental retardation and neuropsychiatric disease in human beings.

Frontal lobe

Anterior regions of the cerebral cortex believed to be involved in planning and higher order thinking.

Functional imaging

Represents a range of measurement techniques in which the aim is to extract quantitative information about physiological function.

Fusiform gyrus

A cortical region running along the ventral (bottom) surface of the occipital-temporal lobes associated with visual processes. Functional activity suggests that this area is specialised for visual face processing and visual word forms.

Gyrus
The circular convolutions of the cortex of which each has been given an identifying name.

(cerebral) Hemisphere
One of two sides of the brain classified as "left" and "right".

Hippocampus
A structure of the limbic system implicated in spatial mapping and memory. This part of the brain is also important in processing and storing long-term memories.

Implicit memory
Memories that cannot be retrieved consciously but are activated as part of particular skills or action, and reflect learning a procedure of a pattern, which might be difficult to explicitly verbalize or consciously reflect upon (*i.e.* memory that allows you to engage in a procedure faster the second time, such as tying a shoe).

IQ
A number held to express the relative intelligence of a person originally determined by dividing mental by chronological age and multiplying by 100.

Left-brained thinking
A lay term based on the misconception that higher level thought processes are strictly divided into roles that occur independently in different halves of the brain. Thought to be based on exaggerations of specific findings of left hemisphere specialisations, such as the neural systems that control speaking.

Limbic system
Also known as the "emotional brain". It borders the thalamus and hypothalamus and is made up of many of the deep structures of the brain.

Lobe
Gross areas of the brain sectioned by function (occipital, temporal, parietal and frontal).

MEG
Magnetoencephalography. A non-invasive functional brain imaging technique sensitive to rapid changes in brain activity. Recording devices (SQUIDS) placed near the head are sensitive to small magnetic fluctuations associated with neural activity in the cortex. Responses to events can be traced out on a millisecond time scale with good spatial resolution for those generators to which the technique is sensitive.

Mental imagery
Also known as visualisation. Mental images are created by the brain from memories, imagination, or a combination of both. It is hypothesised that brain areas responsible for perception are also implicated during mental imagery.

MRI
Magnetic Resonance Imaging. A non-invasive technique used to create images of the structures within a living human brain, through the combination of a strong magnetic field and radio frequency pulses.

Multiple intelligences
Term originally coined to more fully explain the different and equally important ways of processing the environment.

Myelination

Process by which nerves are covered by a protective fatty substance. The sheath (myelin) around the nerve fibres acts electrically as a conduit in an electrical system, ensuring that messages sent by nerve fibres are not lost en route.

Myth of three

Also known as the "Myth of the Early Years". This assumption states that only the first three years really matter in altering brain activity and after that the brain is insensitive to change. This could be considered an extreme "critical period" viewpoint.

Neurodegenerative diseases

Disorders of the brain and nervous system leading to brain dysfunction and degeneration including Alzheimer's disease, Parkinson's disease and other neurodegenerative disorders that frequently occur with advancing age.

Neurogenesis

The birth of new cells in the brain, including neurons.

Neuromyth

Misconception generated by a misunderstanding, a misreading or a misquoting of facts scientifically established (by brain research) to make a case for use of brain research, in education and other contexts.

Neuron

Basic building block of the nervous system; specialised cell for integration and transmission of information.

NIRS

Near Infrared Spectroscopy. Non-invasive imaging method which allows measures of the concentrations of deoxygenated haemoglobin in the brain by near-infrared absorption. (near-infrared light at a wavelength between 700 nm and 900 nm can partially penetrate through human tissues).

Occipital lobe

Posterior region of the cerebral cortex receiving visual information.

OT

Optical Topography. Non-invasive trans-cranial imaging method for higher-order brain functions. This method, based on near-infrared spectroscopy, is robust to motion, so that a subject can be tested under natural conditions.

Parietal lobe

Upper middle region of the cerebral cortex involved in many functions such as processing spatial information, body image, orienting to locations, etc.

Perisylvian areas

Cortical regions that are adjacent to the sylvian fissure – a major fissure on the lateral surface of the brain running along the temporal lobe. Periodicity relates to sensitive periods for certain types of learning.

PET

Positron Emission Tomography. A variety of techniques that use positron emitting radionucleides to create an image of brain activity; often blood flow or metabolic activity.

PET produces three-dimensional, coloured images of chemicals or substances functioning within the brain.

Plasticity
Also "brain plasticity". The phenomenon of how the brain changes and learns.

Right-brained thinking
A lay term based on the misconception that higher level thought processes are strictly divided into roles that occur independently in different halves of the brain. Thoughts to be based in exaggerations of specific findings of right hemisphere specialisation in some limited domains.

Science of learning
Term that attempts to provide a label for the type of research possible when cognitive neuroscience research joins with educational research and practice.

Senile plaques
A clear brain pathology associated with Alzheimer's disease. These are clusters of abnormal cell processes surrounding masses of protein.

Sensitive period
Time frame in which a particular biological event is likely to occur best. Scientists have documented sensitive periods for certain types of sensory stimuli (such as vision and speech sounds), and for certain emotional and cognitive experiences (attachment, language exposure). However, there are many mental skills, such as reading, vocabulary size, and the ability to see colour, which do not appear to pass through tight sensitive periods in their development.

Synapse
Specialised junction at which one neuron communicates with another (called "target cell").

Synaptic density
Refers to the number of synapses associated with one neuron. More synapses per neuron are thought to indicate a richer ability of representation and adaptation.

Synaptic pruning
Process in brain development whereby unused synapses (connections among brain cells) are shed. During the pruning phase, experience and environment decide which synapses will be shed and which will be preserved.

Synaptogenesis
Formation of a synapse.

Temporal lobe
Lateral region of the cerebral cortex receiving auditory information.

TMS
Transcranial magnetic stimulation. A procedure in which electrical activity in the brain is influenced by a pulsed magnetic field. Recently, TMS has been used to investigate aspects of cortical processing, including sensory and cognitive functions.

Trans-disciplinarity
Term used to explain the concept of bridging and fusing completely different disciplines resulting in a new discipline with its own conceptual structure, known to extend the borders of the original sciences and disciplines included in its formation.

Index

OECD PUBLICATIONS, 2, rue André-Pascal, 75775 PARIS CEDEX 16
PRINTED IN FRANCE
(91 2002 02 1 P) ISBN 92-64-19734-6 – No. 52535 2002

Lightning Source UK Ltd.
Milton Keynes UK
15 April 2010

152815UK00001BB/6/P